CERAMIC HEAT EXCHANGER CONCEPTS
AND MATERIALS TECHNOLOGY

CERAMIC HEAT EXCHANGER CONCEPTS AND MATERIALS TECHNOLOGY

by

C. Bliem, D.J. Landini, J.F. Whitbeck,
R. Kochan, J.C. Mittl, R. Piscitella,
J. Schafer, A. Snyder, D.J. Wiggins,
J.M. Zabriskie, B.A. Barna, S.P. Henslee,
P.V. Kelsey

EG&G Idaho, Inc.
Idaho Falls, Idaho

J.I. Federer
E.S. Bomar

Oak Ridge National Laboratory
Oak Ridge, Tennessee

NOYES PUBLICATIONS
Park Ridge, New Jersey, U.S.A.

Copyright © 1985 by Noyes Publications
Library of Congress Catalog Card Number 85-4914
ISBN: 0-8155-1030-6
Printed in the United States

Published in the United States of America by
Noyes Publications
Mill Road, Park Ridge, New Jersey 07656

10 9 8 7 6 5 4 3 2 1

Library of Congress Cataloging in Publication Data
Main entry under title:

Ceramic heat exchanger concepts and materials technology.

    Includes bibliographies and index.
    1. Heat exchangers--Design and construction.
2. Heat exchangers--Materials. 3. Ceramic materials.
I. Bliem, C.
TJ263.C46   1985   621.402'5   85-4914
ISBN 0-8155-1030-6

Foreword

Advanced heat exchanger concepts employing ceramic materials technology, for use in waste heat recuperation from high temperature streams, are described in this book. An estimated 1 to 2 quads of energy are lost each year from high temperature, high fouling, corrosive, industrial waste streams. This is equivalent to the power produced by 14 to 28 1000-MW power plants. Much of this energy is lost from the aluminum, steel, and glass-making industries. Waste energy from these operations is among the most difficult to recover because the waste energy streams are: (a) high temperature [flue gas temperatures range up to 1650°C (3000°F)]; (b) high fouling (particulates build up on heat exchanger surfaces); (c) corrosive to metallic alloys, especially at desired preheat temperatures of 1100°C (2000°F) or higher.

Only ceramic materials have sufficient high temperature strength and oxidation resistance to serve as primary heat exchanger elements providing air preheated to 1100°C. Ceramic materials are being considered for use in other high temperature applications such as in automobile engines and turbochargers. In waste heat recovery systems, ceramics are being formed into tubes or used in granular form in fluidized bed/pebble bed applications.

In an effort to focus on waste heat recovery from high temperature, dirty, industrial exhaust streams this book conceptually develops, evaluates, and compares potential new concepts with emerging and state-of-the-art concepts in terms of corrosion/durability, fouling, performance, operation and maintenance, and economics.

Part I discusses the design, functional and cost requirements for high-temperature, heat recovery systems (recuperators) and describes the state-of-the-art systems, emerging industrial technologies and new concepts. All systems/concepts are then evaluated and compared with respect to corrosion/durability, fouling, performance, operation and maintenance, and economics.

Part II assesses the material property data and related test method, life prediction, and product reliability technology available for ceramic materials used in heat exchanger designs intended to conserve energy by preheating industrial furnace combustion air. Several candidate ceramic materials are discussed that have the potential to withstand the high temperatures and corrosive constituents of waste heat streams from industrial processes such as aluminum remelt, steel, and glass-making furnaces where metallic recuperators have problems. Currently available and emerging heat exchanger designs are reviewed and their operational problems are considered. In addition, Part II identifies areas in which more information is needed.

The information in the book is from:

Evaluation of Selected Advanced Heat Exchangers for Waste Heat Recuperation of High Temperature Streams by C. Bliem, R. Kochan, J. Mittl, R. Piscitella, J. Schafer, A. Snyder, D. Wiggins and J. Zabriskie of EG&G Idaho, Inc. for the U.S. Department of Energy, February 1984.

An Assessment of Ceramic Materials Technology for Heat Exchangers by B.A. Barna, S.P. Henslee, P.V. Kelsey, D.J. Landini, J.C. Mittl, J.F. Whitbeck, D.J. Wiggins and J.M. Zabriskie of EG&G Idaho, Inc. and J.I. Federer and E.S. Bomar of Oak Ridge National Laboratory for the U.S. Department of Energy, January 1984.

The table of contents is organized in such a way as to serve as a subject index and provides easy access to the information contained in the book.

Advanced composition and production methods developed by Noyes Publications are employed to bring this durably bound book to you in a minimum of time. Special techniques are used to close the gap between "manuscript" and "completed book." In order to keep the price of the book to a reasonable level, it has been partially reproduced by photo-offset directly from the original reports and the cost saving passed on to the reader. Due to this method of publishing, certain portions of the book may be less legible than desired.

NOTICE

The materials in this book were prepared as accounts of work sponsored by the U.S. Department of Energy. Publication does not signify that the contents necessarily reflect the views and policies of the contracting agencies or the publisher, nor does mention of trade names or commercial products constitute endorsement or recommendation for use.

Contents and Subject Index

PART I
CERAMIC HEAT EXCHANGER CONCEPTS

1. INTRODUCTION 2
   Purpose of Report 2
   Background 3
   References 5

2. DESIGN/FUNCTIONAL/COST REQUIREMENTS 6
   Waste Heat Stream Description and Design Information 6
      Glass Melting Furnaces 7
      Aluminum Melting Furnaces 8
      Steel Soaking Pit 9
   General Design, Functional, and Cost Requirements 10
      Performance 11
      Durability 12
      Cost of the Heat Exchanger Component 12
   Reference 12

3. RECUPERATOR DESIGNS AND APPLICATIONS 13
   Description of State-of-the-Art Recuperators and Emerging
      Technologies 13
      State-of-the-Art-Recuperators 13
      Emerging Technology Recuperators .16
   Description of EG&G Concepts 20
      Concept Development 24
      Concept Selection 24
         Basic Requirements Filter 28
         Thermal-Hydraulics Filter 28

viii   Contents and Subject Index

       Materials Filter 28
       Fabrication Filter 28
       Filtering Results 28
    Detailed Description of EG&G Concepts 33
       Phase-Change Shot Tower 33
       Ceramic Fiber Heat Exchanger Concepts 38
          Woven Tube Ceramic Heat Exchanger 38
          Ceramic Paper Heat Exchanger.....................45
    Detailed Description of Ceramic Heat Pipe Concept 58
  References...63

4. EVALUATIONS 65
  Waste Streams Corrosive Effect on Material Durability 65
    Heat Exchanger Ceramic Corrosion Evaluation 66
       Silicon Carbide (SiC) 67
       Cordierite 69
       Refractory Brick Compositions 69
       Alumina (Al_2O_3)..............................70
       Zirconium Oxide (ZrO_2)..........................70
    Summary of Corrosion Evaluation of Ceramic Materials 71
    References....................................72
  A Comparison of the Performance of Recuperators 73
    Recuperator Fouling Considerations 73
       Fouling Potential in Glass Melting Furnaces 74
       Fouling Potential in Steel Soaking Pits.................75
       Fouling Potential in Aluminum Remelt Furnaces............76
       Fouling Summary 76
       Prevention of Fouling Problems 77
       Evaluation of Fouling in Recuperator Systems 77
    Maintenance...................................79
       Preventive Maintenance 80
       Annual Maintenance 80
       Maintenance Evaluation 81
    A Comparison of the Thermal Performance of Recuperator
    Furnace Systems................................84
       Recuperator Process Conditions for Each Application 84
       Fuel Saving and Performance Data 88
       Discussion of Results 94
    Operating Considerations 98
       Recuperator Control Systems 98
       Operating Labor 98
       Operating Utilities.............................102
       Operational Evaluation..........................102
  Economic Analysis 102
    Cost and Energy Savings Estimates 104
    Economic Analysis Methodology 107
    Sensitivity of Economic Results to Input Parameters 113

Variation in Economic Parameters with Variation in Selected
  Design Assumptions... 113
Potential Market.. 117
Conclusions... 118
Summary of Evaluation.. 119
Concept Evaluation/Comparison for the Steel Soaking Pit
  Furnace Applications.. 122
Concept Evaluation/Comparison for the Aluminum Remelting
  Furnace Applications.. 123
Concept Evaluation/Comparison for the Glass Melt Furnace
  Applications.. 123
Conclusions of Evaluation/Comparisons............................. 124

5. RESEARCH AND DEVELOPMENT NEEDS 125
Fouling... 125
Corrosion... 126
Economics... 126
Performance... 126
Woven Tube Heat Exchanger Concept................................ 126
Paper Finned Plate Concept....................................... 127

6. REFERENCES 128

APPENDIX A: DRAWINGS AND DESCRIPTIONS OF MATERIALS CONCEPTS 129

APPENDIX B: DRAWINGS AND DESCRIPTIONS OF FABRICATION CONCEPTS 133

APPENDIX C: DRAWINGS AND DESCRIPTIONS OF COMPONENT CONCEPTS 140

APPENDIX D: DESCRIPTIONS OF PROCESS CONCEPTS WITH LISTS OF POTENTIAL ADVANTAGES AND DISADVANTAGES............ 149

APPENDIX E: PHASE SHOT TOWER DESIGN CALCULATIONS 171

APPENDIX F: WOVEN TUBE AND PAPER FINNED PLATE DESIGN CALCULATIONS 200

APPENDIX G: STATE-OF-THE-ART AND EMERGING CONCEPTS COSTS 218

PART II
MATERIALS TECHNOLOGY

1. INTRODUCTION 226

2. CURRENT CERAMIC HEAT EXCHANGER DESIGNS 228
State-of-the-Art Designs 228
GTE Recuperator 228
Hague International Recuperator 233
Emerging Designs 235
Garrett-AiResearch Design 236
Babcock & Wilcox Design 236
Uintex/Millcreek Design 236
Solar Turbines International Design 240
Midland-Ross Design 240
Conclusions 243

3. ASSESSMENT OF CERAMIC RECUPERATOR DESIGN NEEDS 244
Design Process 244
Design Requirements 245
Material Data Requirements 247
Conclusions 249

4. CERAMIC MATERIALS FOR HIGH TEMPERATURE HEAT EXCHANGER APPLICATION 251
Selection of Candidate Materials 251
Description of Candidate Silicon Carbides 253
Sintered-α and Sintered-β SiC 253
Recrystallized SiC 253
Reaction Sintered or Siliconized SiC 253
Nitride-Bonded SiC 254
Oxide-Bonded SiC 254
Chemical Vapor Deposited SiC 254
Available Mechanical Properties 255
Properties Available from Manufacturers 255
Properties Available from the Literature 257
Properties of Materials Exposed to Corrosive Environments 267
Steel Soaking Pit Exposure 268
Aluminum Remelt Exposure 272
Glass Furnace Exposure 273
Residual Oil Combustion Gases Exposure 277
Coal Oil Mixture Combustion Exposure 279
Results of Exposure to Acidic Ash 279
Results of Exposure to Basic Ash 282
Summary 284
Conclusions 285
Improved Materials 286
Improved Processes 286
Synthesis of SiC Powder 286
Gel Process 290
Polymeric Precursor Process 290
Laser-Heated Gas-Phase Reaction Process 293
Toughened Ceramics 293

   Aluminum Oxide............................293
   SiC and Si_3N_4............................294

5. DESIGN-RELATED MATERIAL TECHNOLOGIES............295
  Mechanical Strength Test Methods........................295
   Tensile Test Methods...............................295
   Bend Test Methods................................299
   Radial Compression Test Methods......................299
   "C" or Split Ring Test Methods......................304
  Time-Dependent Testing...............................306
   Creep Test Methods...............................306
   Slow Crack Growth Test Methods......................307
   Corrosion Test Methods............................310
   Conclusions.....................................311
  Lifetime Prediction Techniques.........................312
   Fracture Mechanics Approaches.......................312
    Statistical Analysis............................313
    Proof Testing.................................320
    Proof Testing: Applications......................322
   Stress-Rupture Testing............................325
   Conclusions.....................................327
  Product Reliability Techniques.........................328
   NDE..328
    Physics of Ceramics NDE........................328
    Currently Applied Technology in Production.............329
    Technology Under Development....................330
     Ultrasonic Scanning..........................331
     Scanning Photo-Acoustic Microscopy................331
     Pulse-Echo Acoustic Microscopy...................332
     Microfocus X-Ray...........................333
     Acoustic Microscopy.........................333
     Conventional Ultrasonics......................336
     Acoustic Emission (AE).......................336
     Other NDE Methods.........................337
    Cost Effectiveness of NDE.......................337
   Proof Testing...................................339
   Assessment and Future Objectives.....................340

6. CONCLUSIONS AND RECOMMENDATIONS..................342

7. REFERENCES...345

APPENDIX A: CERAMIC MATERIALS PROPERTIES DATA.........352

APPENDIX B: SI CONVERSION FACTORS......................385

Part I

Ceramic Heat Exchanger Concepts

The information in Part I is from *Evaluation of Selected Advanced Heat Exchangers for Waste Heat Recuperation of High Temperature Streams* prepared by C. Bliem, R. Kochan, J. Mittl, R. Piscitella, J. Schafer, A. Snyder, D. Wiggins and J. Zabriskie of EG&G Idaho, Inc. for the U.S. Department of Energy, February 1984.

1. Introduction

1.1 Purpose of Report

The purpose of this report is to show the development of some appropriate heat exchanger concepts and give a comparison and evaluation with existing concepts. This activity will act as a forerunner to the solicitation of concepts from industry and the evaluation of those concepts. One or more of the concepts from this study and hopefully some concepts from industry will then be further developed and evaluated and eventually tested in a module size in the actual or simulated environments of interest. The testing will occur in conjunction with other applicable program technological developments.

Since the heat exchanger (in this report) is the critical component for recovering and using waste heat, it is the focal point for most of the Department of Energy's Industrial High Temperature Waste Heat Recovery (IHTHR) program activities. Heat exchanger designs need to be developed specifically for construction with ceramic materials because of their special properties. Other technology developments in fabrication techniques, joining techniques, new materials and coatings can then be tested in the newly developed heat exchanger designs. Corrosion/fouling, and destructive and nondestructively obtained material properties can also be verified particularly in reference to the durability and performance of the design.

Specifically, this report includes design, functional, and cost requirements of a heat exchanger for high-temperature corrosive environments, a description of the state-of-the-art and emerging high-temperature heat exchanger technologies, and how additional heat exchanger concepts were obtained, selected, and developed. The newly developed concepts are then evaluated and compared with the state-of-the-art and emerging technology concepts in terms of corrosion/durability, fouling, performance, operation and maintenance, economics, and technological contribution. Finally recommendations and research and development needs are discussed.

1.2 Background

The United States currently purchases approximately 73×10^{15} Btu/yr of fuels and electricity. Approximately one-third of these fuels and electricity are purchased by the industrial sector. The Environmental Protection Agency (EPA) estimates that 23% of the major or energy intensive industrial fuels and electricity energy consumption is discharged as waste heat in flue gases.[1] For any single high temperature (>1500°F), direct heating process, as much as 67% of furnace energy is lost in the flue gas. The higher the furnace input temperature, the greater the percentage of energy lost.[2] The recovery of waste heat from industrial high-temperature flue gases could potentially save one to two quadrillion Btu/yr within the United States alone.

Industry has attempted to decrease the amount of this lost energy with waste heat recovery systems (recuperators or regenerators). However, because of the high temperatures and sometimes corrosive constituents in the waste heat streams, durability of the construction materials has limited recovery. Recovery of waste heat from high-temperature industrial gas streams is most commonly done by preheating combustion air using either a recuperator or a regenerator. Generally, a recuperator uses the exhaust gas to directly, or through a partition wall, heat the combustion air. A regenerator normally allows hot exhaust gas and combustion air to move alternately through the same passage, thus indirectly heating the combustion air. Historically, recuperators and waste heat boilers have been constructed of metal. However, these conventional metallic heat exchangers cannot survive for extended periods in high-temperature, dirty environments without incurring severe performance penalties. Failure of the system impacts not only its ability to recover waste heat, but the operation of the process to which it is attached as well. Corrosion has debilitating effects on most all materials, causing premature failures and/or excessive leakages. Fouling decreases heat transfer rate, increases pressure drop, and adds expense due to increased surface area and the necessity for cleaning and refurbishing.

Ceramics, an alternative to metals, offer the potential to resist high-temperature, corrosive environments by allowing significantly higher material temperatures and by offering resistance to many corrosive constituents in industrial waste streams. However, certain technical limitations exist which severely restrict the application of advanced ceramic heat exchangers in high temperature fouling and corrosive waste streams. These include the high costs for ceramic fabrication, problems with satisfactorily joining ceramics, the lack of data and accurate methods to predict thermal-structural behavior of ceramics especially as affected by long-term exposure to corrosive environments, the sensitivity of some of the more corrosive-resistant ceramics to thermal shock fracture, and the inability to detect and evaluate flaws causing failure. These limitations are more specifically addressed in a soon to be published report, "An Assessment of Ceramic Materials Technology for Heat Exchangers."[3]

The Department of Energy has realized the above-stated need and has developed an Industrial High-Temperature Waste Heat Recovery (IHTHR) Program for developing a technology base in critical areas of high-temperature, waste heat recovery that reduces the technical risk to a level that will motivate the private sector to continue developing and implementing the necessary technology. The IHTHR Program focuses on waste heat recovery from high-temperature, dirty industrial exhaust streams such as from steel soaking pits, aluminum remelt furnaces, and glass melting furnaces. The previously addressed technology limitations for the application of ceramics to high-temperature heat exchangers are specifically addressed in this program.

References

1. S. R. Latour and J. G. Menningmann, *Waste Heat Recovery Potential in Selected Industries*, Contract No. 68-01-4454, DSS Engineers, Inc., Ft. Lauderdale, Florida, February 1981.

2. J. J. Cleveland, J. M. Gonzalez, K. H. Kohnken, *Ceramic Heat Recuperators for Industrial Heat Recovery*, DOE/CS-40174-T2, August 1980.

3. D. J. Wiggins (ed.), *An Assessment of Ceramic Materials Technology for Heat Exchangers*, EGG-SE-6367, (to be published in November 1983).

2. Design/Functional/Cost Requirements

To develop waste heat recovery design concepts, it is necessary to define design, functional, and cost requirements. Thus, it is necessary to understand the process constraints and environments to which the waste heat recovery systems will be exposed. In a recently published report on high-temperature waste heat stream characterization,[1] process constraints and waste stream constituents are addressed for three selected waste heat streams: steel soaking pits, glass melting, and aluminum remelting. These streams will be the focus of further design considerations in the IHTHR Program and in this report.

Once the process constraints and environments are understood, the problems with using ceramic waste heat recovery systems or recuperators in these processes need to be considered so that requirements can address these problems and the concepts can be evaluated accordingly.

The following section (Section 2.1) briefly summarizes available design information on waste heat streams of interest. Section 2.2 recognizes industry's concerns in design, function, and cost of high temperature recuperators and develops the requirements necessary to obtain greater industry acceptance and use of newly developed concepts. These requirements form the basis for evaluation of existing and newly developed concepts presented in Section 4.

2.1 Waste Heat Stream Description and Design Information

The information available for describing a typical furnace or for characterizing the waste gas streams of glass melt furnaces, aluminum remelt furnaces, or soaking pits is limited. Thus, information from Department of Energy contractors on specific furnaces was used as a guide. Each process furnace has slightly different operating parameters which will affect recuperator design. These are discussed in the following sections. All the process furnaces generally operate though, at the same near-atmospheric pressures, thus requiring that the recuperator design have very little gas side pressure drop unless a fan or ejector is used.

2.1.1 Glass Melting Furnaces

Glass melting furnaces are generally continuous furnaces at constant temperatures ranging between 2800°F and 3000°F depending on the feedstock composition. A constant temperature of 2800°F is assumed for this report.

The flow rate of the gases ranges from 440,000 to 1,200,000 scfh. These numbers are for a continuous furnace and are determined by the production size (furnace size). The major components of the gas are the combustion products of the fuel. The critical corrosion/fouling substances are introduced into the gas stream usually as fuel contaminants and batch carryover and include Al, Ca, Cl, F, K, Mg, Na, S, Si, and V. The glass melting process can be carried out in one of three types of melting tanks: regenerative furnace, day pot, or day tanks.

Most glass tonnage is melted in continuously operating regenerative or recuperative furnaces. The regenerative furnaces utilize two chambers of refractory (checker work). While the combustion flue gases heat the refractory in one checker work, the other checker work preheats air. The operation is reversed every 10 to 30 minutes. These furnaces maintain a larger production rate than do the recuperative furnaces which have one continuously operating tube and shell heat exchanger to preheat the combustion air. Regenerative type furnaces are divided into two classes--side port and end port, depending on the firing pattern and the type of glass being produced.

For smaller batches, such as the production of special compositions of glass, the day pot or day tank is used. Day pot capacities range from 20 pounds to 2 tons while the day tanks are larger. The major difference between the two is the material of construction used for the vessel walls. The day tanks and pots are used to melt 24-hour batches.

Although the use of an advanced recuperator on a day tank or day pot may be most realistic because of its ready adaptability, a complete replacement of the checker works of a regenerative furnace by a recuperator which can be placed in a 4-ft diameter by 4-ft length ducting is assumed for the purposes of this report. A high flow rate of approximately 57,000 lb/h (including the additional preheated air which flows through the burner) of exhaust gas is also used.

2.1.2 <u>Aluminum Melting Furnaces</u>

The process from electrolytic reduction, inclusive through remelting, alloying and the final casting of the aluminum, is described here as the primary aluminum process. Following electrolytic reduction (smelting) the molten aluminum is siphoned from the reduction pot directly into a holding furnace or remelt furnace. Here it is alloyed, degassed and cast into an ingot of the proper metallurgical structure and size for metal working. The molten aluminum may be cast into ingots prior to remelt. The primary aluminum process may include the remelt of scrap from rolling and machining of the aluminum metal. Recycling is the secondary aluminum process. The recycled aluminum is treated in the same manner as the new ingots. It, however, is contaminated with other metals such as stainless steels, alloying agents, and dirt (hydrocarbons from oily residues and paints).

Remelting of aluminum generally takes place in a continuous reverberatory furnace--a furnace where the flame is developed some distance above the hearth and heat is reflected onto the hearth by an arched or sloped roof. In order to burn off any hydrocarbons or volatiles and to bring the metal up to temperature, the ingot or baled scrap is allowed to remain on the forehearth before being pushed into the bath of molten metal. Following melting and the addition of alloying agents, the molten metal is skimmed to remove oxide and dross from the surface. The metal is run to a holding furnace, generally also a reverberatory furnace, and the temperature is adjusted to the temperature for casting. Fluxing and degassing agents are added (the steps of the remelt and holding furnace may be combined in one furnace) and there may be a quiescent period where particulate matter is allowed to settle. The metal is then cast by a semicontinuous or direct-chill process.

The reverberatory remelt furnaces of the aluminum industry nominally operate at 2000 to 2800°F. A temperature of 2100°F is assumed for this

Design/Functional/Cost Requirements   9

report with maximum thermal transients of 250°F/h. A space of 4 x 4 x 4 ft in the ducting is assumed for the recuperator. The average gas flow rate varies depending on furnace capacity and fuel input but is assumed to be 9990 lb/h (including the additional preheat air which flows through the burner) at high fire with a 5:1 turndown ratio. Fluxes, degassing agents, and the batch composition will contribute to the corrosive/fouling components of the waste gas stream; included are Al, Ca, Cl, Fe, K, Mg, Na, S, F, and V.

2.1.3 **Steel Soaking Pit**

In order to produce a metal of uniform temperature for rolling and shaping steel finishing, soaking pits and billet reheat furnaces are used. Following pouring, the solidified ingot is stripped from its mold and placed in a soaking pit where it is brought up to the temperature required for rolling to produce slabs, blooms or billets. These are subsequently heated in a reheat furnace for further shaping and treating of the steel to produce the finished steel product. With continuous casting the soaking pits may be bypassed.

There are six major types of soaking pits: one-way fired, bottom center-fired, circular, bottom two-way fired, top two-way fired, and electric. The recuperative soaking pit is the oldest of the modern pits. The recuperative one-way fired pit where the burner is continuously fired in a combustion space above the ingots is one of the most used types of soaking pits.

The temperature of steel soaking pits ranges from 1300 to 2450°F, but is assumed for this report to be 2450°F with maximum thermal transients of 400°F/h. This temperature depends on the firing rate of the furnace with lower firing (low air and fuel flow) conditions being used to maintain the higher temperatures. Approximately half of the fired time is at low-fire or near low-fire conditions. The temperature is also determined by the type of alloys being heated or soaked. Soaking pits are subjected to large changes in temperature due to the addition or removal of materials. The waste gas flow of the soaking pit and the reheat furnace is dependent on the firing rate of both. A flow of 250,000 scfh corresponds to an average of the maximum fire rate of all the furnaces. The flow may decrease to 50,000 scfh when the burner is operated at low fire conditions. A flow rate of 9114 lb/h (including the additional preheated air which flows through the burner) at high fire with a turndown ratio of 4:1 is assumed for this report. Fuel contaminants, topping compounds, oxidation of the metal surface, and furnace floor constituents will contribute to the corrosive constituents of the waste gas stream; included are Ca, Fe, K, Na, S, and Mg.

2.2 General Design, Functional, and Cost Requirements

There are several basic requirements or design criteria which a new advanced waste heat recovery system must meet to be equivalent or better than current state-of-the-art technology and to satisfy the IHTHR's overall program objectives:

- The materials of construction must be compatible with the waste stream constituents of the aluminum remelting furnace, the glass melting furnace, and/or the steel soaking pit

- The heat exchanger must be able to operate at material temperatures greater than 1900°F up to 3000°F depending on the waste heat stream and design of the heat exchanger

- The heat exchanger must be capable of eventually being fabricated in a commercialized unit, e.g., the concept must have the potential to be easily scaled from laboratory size to commercial size

- As part of DOE's role, the system must be ready for further industrial development by the private sector within 5 to 10 years such that the concept is not so innovative that it would take years to develop or so near term that industry can develop it themselves.

Besides improving on the state-of-the-art technology and satisfying DOE goals, a major requirement is the acceptance of the system by industry. Industry is mainly concerned with the economics of retrofitting a recuperator to an existing steel soaking, an aluminum melting furnace, or a glass melting furnace. This concern can mainly be expressed in terms of return on investment, installation, and maintenance and operational impacts. Common industry concerns include:

- How much are the waste heat recovery units going to cost

- How much are they going to save in fuel

- How much are they going to save in productivity

- How much will it cost to install them in terms of labor and process downtime

- How much additional money and labor will it cost in maintaining and operating them

- How long will they last before they have to be replaced

- How reliably will they continue to operate?

Although all these issues must be addressed in detail to satisfy industry's economic concerns, they can be summarized into three critical areas to be addressed during the development and design of new advanced waste heat recovery systems and compared with existing systems: the performance of the waste heat recovery system, durability of the heat exchanger component, and the cost of the heat exchanger component. A more detailed discussion of the requirements for these three critical factors follows.

2.2.1 Performance

Most of the furnaces discussed in this paper operate over a wide range of temperature. Therefore, an advanced recuperator system including heat exchanger, fans, connecting ducts, and controls must save energy over the expected operating fluctuations. The system configuration and materials should be selected to mitigate fouling so that the system continues to save energy over its entire operating lifetime. The system should be designed for use on more than one furnace and/or process application. Installation of the system should be able to be accomplished without much disruption of plant operation. Finally, the system should be operated and maintained with little impact on plant productivity, it should be easily learned and controlled by the operator, and it should have only a minimum environmental impact.

2.2.2 Durability

An advanced waste heat recovery system must last a reasonable length of time with routine maintenance, depending on system design and economics. This requires that the configuration and/or material not only resist corrosion sufficiently or accommodate a certain amount of corrosive degradation, but be designed to withstand mechanical loads and thermal stresses caused during normal or off-normal design conditions. The configuration and/or material must be easily maintained with minimal replacement of costly parts.

2.2.3 Cost of the Heat Exchanger Component

An advanced waste heat recovery system must be capable of cost-effective fabrication into the desired configurations as reflected in material costs and fabrication costs.

Reference

1. P. M. Wikoff, et al., "High Temperature Waste Stream Selection and Characterization," EGG-SE-6349, August 1983.

3. Recuperator Designs and Applications

This section provides design descriptions of each recuperator concept evaluated in this report. Included are two commercially available units or state-of-the-art concepts, five emerging concepts, and three "new-idea" concepts developed by EG&G Idaho. These concepts are evaluated and their advantages and disadvantages discussed later in this report in Section 4. The materials used in these units are also described as part of the evaluation in more detail in Section 4.

3.1 Description of State-of-the-Art Recuperators and Emerging Technologies

Seven ceramic recuperator concepts with the potential of operating in the selected high temperature waste heat streams have been or are being developed. Two of these are commercially available and are considered state-of-the-art. The other five are classified as emerging technologies and are currently undergoing prototype development and testing. These five concepts are discussed in more detail in the following paragraphs and are summarized in Table 3.1.

3.1.1 State-of-the-Art-Recuperators

Two currently available ceramic recuperators are a finned plate heat exchanger manufactured by GTE Sylvania and a cross counterflow tubular heat exchanger manufactured by Hague International. These ceramic units offer performance advantages over conventional metallic units and are capable of producing 322 to 1090 K (120 to 1500°F) preheated combustion air (i.e., heat exchanger effectiveness up to 50%).

The "Super Recuper"^{R}, the finned plate heat exchanger manufactured by GTE Sylvania, is very compact. The basic module is a cross-flow matrix that can operate in a counter flow configuration when staged (see Figure 3.1). The modules are made of extruded plates that are stacked and bonded with a proprietary compound before firing. Various sizes and

TABLE 3.1 SUMMARY OF DEVELOPED STATE-OF-THE-ART AND EMERGING TECHNOLOGIES

Heat Exchanger Type	Configuration	Materials	Unique Features	Class Status	Company Developing Concept
Finned Plate	Cross-flow Matrix Counterflow When Staged	Cordierite MAS 8400	Crosswise Staked & Bonded Extruded Plates; Spring Loaded Housing and Seals; Various Sizes and Passages	SOTA Commercial	GTE
Tube In Shell	Cross-flow Single and Two Pass Tube Bundles	Phosphate Bonded SiC HT 25 (2600°F) HT 35 (2800°F)	Spring Loaded Mounts and Seals; Radially Finned and Smooth Tubing	SOTA Commercial	Hague
	Cross-flow Single Counterflow When Staged	Si N Bonded SiC - Tubesheet and Enclosures SiC Tubes	Compression Loaded Structure; Cement and Compliant Seals	Emerging Technologies Prototype Developing	Garrett-AiResearch/ GRI (HRBDR)
	Axial Counterflow	Sintered SiC NC 430	Viscous Glass Seals; Braze Bonded Tubes Ball and Socket Joints; High Pressure - 100 PSI	Emerging Technologies Prototype Developing	Solar Turbines
Tube In Tube	Cross-flow Past Annular Counterflow	SiC - Tubes Inner: CS 101 Outer: NC 430 Tubesheets	Hanging Bayonet Mount Insulated and Cooled Tubesheets; Ceramic Paper for Seals	Emerging Technologies Prototype Developing	B&W (HTBDR)
Helical	Counterflow	Alumina Chromia - ECP - 3 Magnesia Chromia Spinel - X - 81 Unichrome	Segmented-Stackable; Compression Gaskets; Single and Double Helixes, Housing; Labyrinth Seals With and Without Fins	Emerging Technologies Prototype Developing	Uintex/Mill- creek Glass
Heat Wheel	Segmented Matrix	Mas - Matrix Thermo Sil 120 Castable Interam - Joints	Matrix Segments Cast In Place; Resilient Bonding to Metal Housing; Labyrinth Seals	Emerging Technologies Prototype Developing	Midland - Ross (GRI)

Recuperator Designs and Applications 15

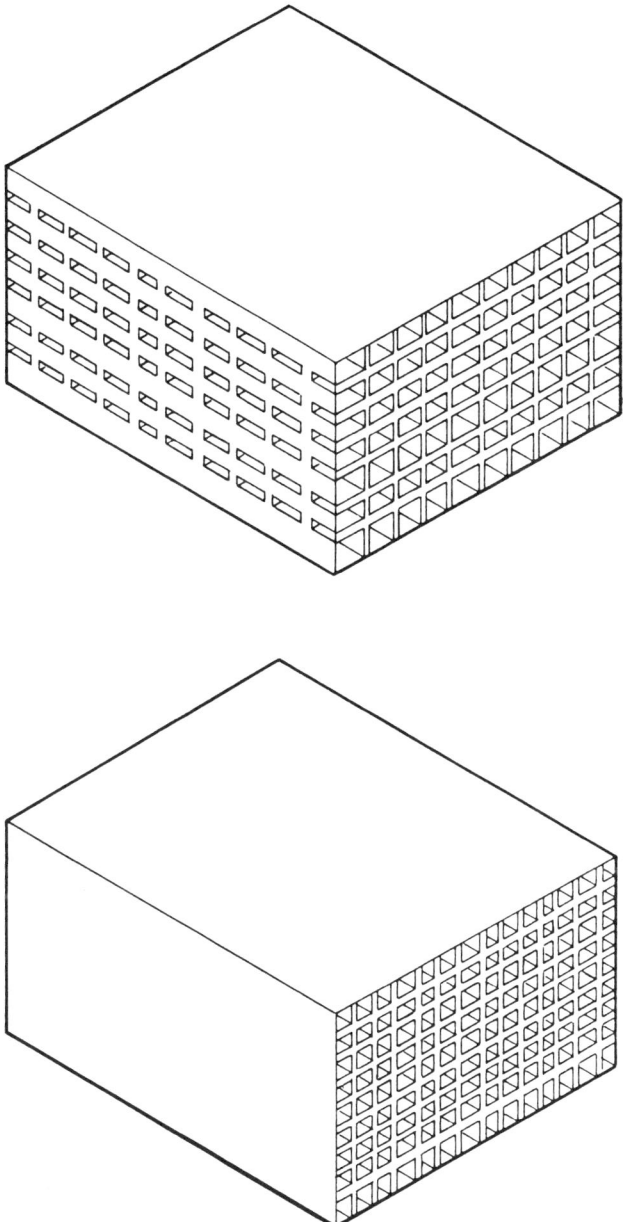

Figure 3.1 Cordierite finned plate cross-flow recuperator and regenerator configuration.

passages are available. The material is cordierite, a magnesia alumina silicate (MAS). For the extreme environment of borosilicate glass furnaces some experiments using fluxed zirconia silicate and alumina coated MAS were performed.[1,2]

The "Cerhx"[R] recuperator manufactured by Hague International, the other commercially available ceramic heat exchanger, is a basic tube-in-shell design using two passes of cross-flow tube bundles (see Figure 3.2).[3] The tubes are phosphate-bonded silicon carbide (SiC) and are available in plain and finned versions. The tubes use spring-loaded mounts and seals.

3.1.2 Emerging Technology Recuperators

The Department of Energy is currently funding two companies (Garrett-AiResearch Co. and Babcock & Wilcox) to develop two high-temperature burner duct recuperator (HTBDR) systems using a technology base that is somewhat advanced compared to existing ceramic heat exchangers. Although the predicted performance of these two recuperators is high [i.e., an effectiveness of approximately 80% and a preheat combustion air temperature up to 1366 K (2000°F)], their technical risk is high and durability is unknown. This is attributable to a very limited design data base for ceramic materials for waste heat recovery applications.

The Garrett-AiResearch design is a tube-in-shell exchanger (see Figure 3.3).[4] It is divided into two sections--ceramic and metallic. The ceramic section uses two cross-flow passes through SiC tubes. The tube sheet and enclosures are silicon nitride bonded SiC and everything is held in place by a system of spring-loaded tie rods, square tubes and crossbeam channels mostly constructed of RA330, a high temperature alloy.

The other HTBDR design is by Babcock & Wilcox and features a tube-in-tube configuration (see Figure 3.4).[5] Flue gas flows past the outside of the hanging bayonets. Preheated air flows through the bayonets via the inside of an inner tube and then back up an annulus formed by the outer and inner concentric tubes. Both inner and outer tubes are SiC with

Recuperator Designs and Applications 17

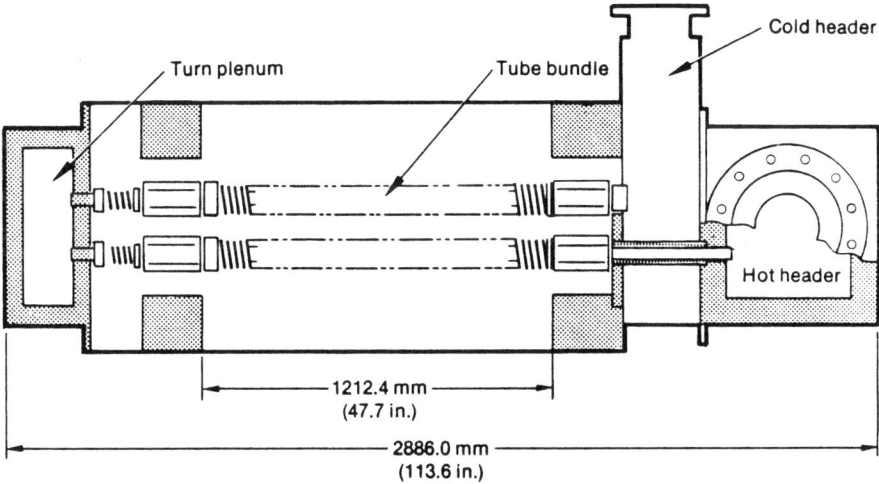

Figure 3.2 Two-pass ceramic spring loaded tube-in-shell heat exchanger.

18 Ceramic Heat Exchanger Concepts and Materials Technology

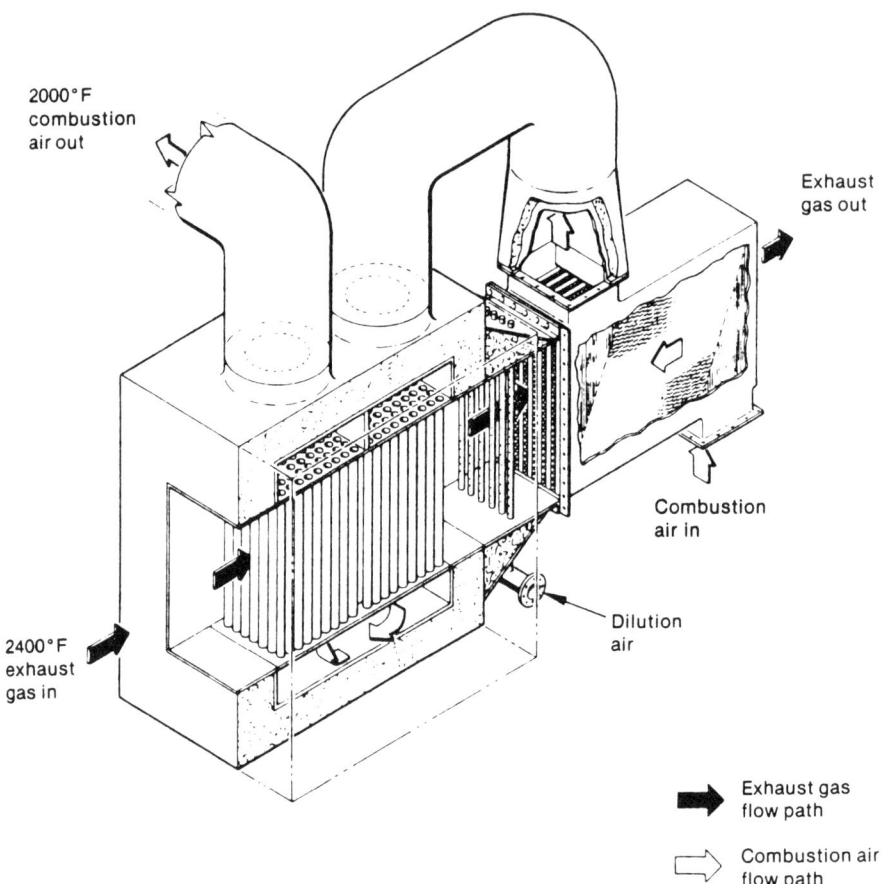

Figure 3.3 SiC compliant seal tube-in-shell recuperator.

Recuperator Designs and Applications 19

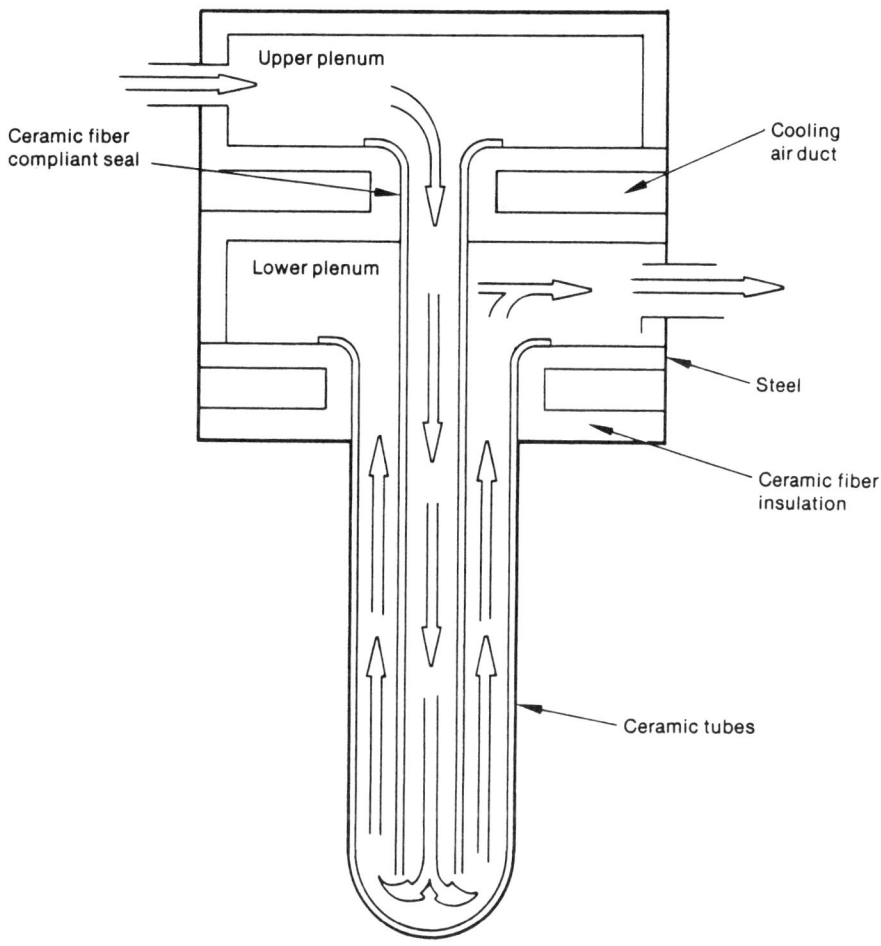

Figure 3.4 SiC tube-in-tube recuperator.

the inner tubes being CS 101 and the outer tubes being NC 430 (these materials have changed recently, but the new materials will not be considered for this evaluation). All the tubes hang from the cooled 309 SST tube sheets. Tube to tubesheet seals are compliant ceramic paper.

Unitex/Millcreek designed and tested a recuperator for use in the harsh environment of a glass furnace.[7] It is a helical counter flow model which uses stackable modules (see Figure 3.5). Alumina chromia (ECP-3) and magnesia chromia spinal (X-81 and Unichrome) provide potentially good performance in the soda-lime glass environment.

Solar Turbines has conceptualized a tube-in-shell design that is unique in that it is designed to operate at much higher pressures than the previously mentioned designs: i.e., approximately 100 psig vs. atmospheric (see Figure 3.6).[7,8] This heat exchanger is designed to be used for recuperating heat from gas turbine exhaust streams. The tubes are manufactured from sintered SiC (NC 430) in 4 to 8 ft segments which are brazed together (with an alignment collar) to form 15 to 20 ft tubes. Airflow through the tubes is in an axial counter flow configuration and is ducted into and out of the exhaust stream by means of headers. The headers at the cold end of the tubes have a relaxing joint. At first, a viscous glass seal was tried, but tended to extrude from the joints at operating pressures. The next version used spherical ball and socket joints to allow for thermal expansion and misalignments.

Midland-Ross/GRI has proposed a segmented matrix heat wheel also designed for use with the relatively clean exhaust from gas turbines (see Figure 3.7).[9] This regenerator design uses segmented matrix elements extruded from MAS and cast in place with thermo sil 120^R. Interam^R mat is used as a reliant bonding media between the matrix assembly and the metal housing. This design also features axial labyrinth seals.

3.2 Description of EG&G Concepts

EG&G Idaho has developed three conceptual designs of advanced heat recovery systems to be applied on steel soaking pit, aluminum remelt, or

Recuperator Designs and Applications 21

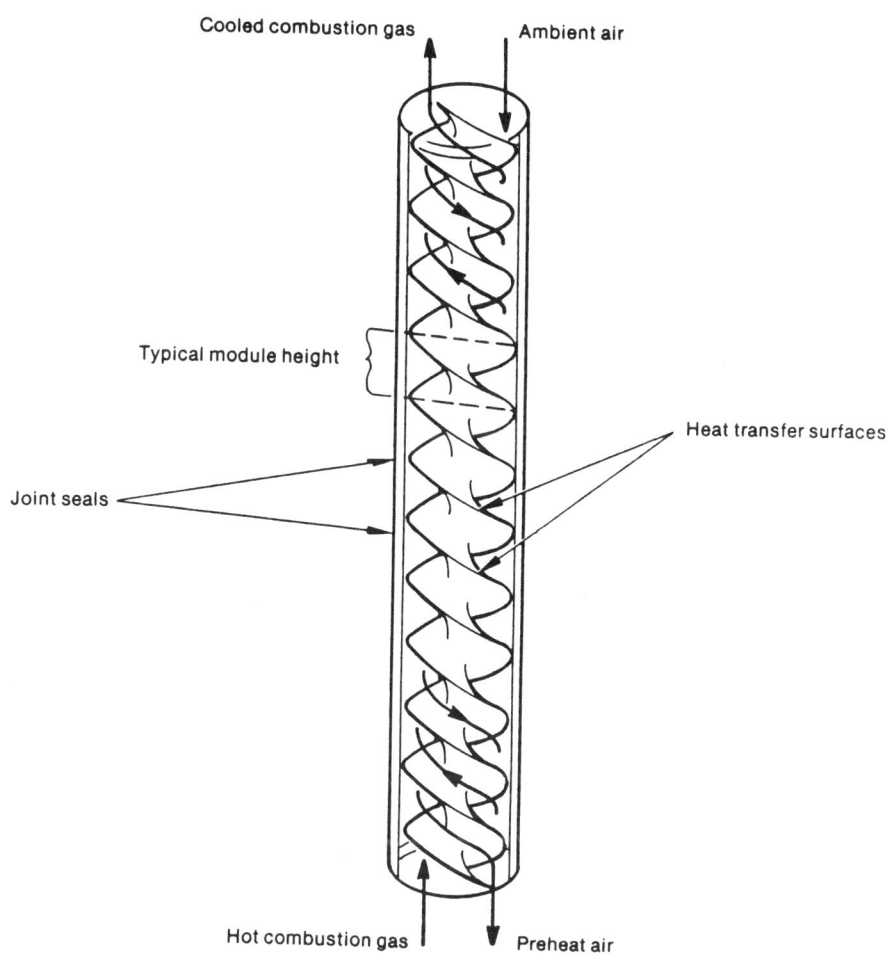

Figure 3.5 Schematic of the modular counterflow recuperator with a helical interface.

22 Ceramic Heat Exchanger Concepts and Materials Technology

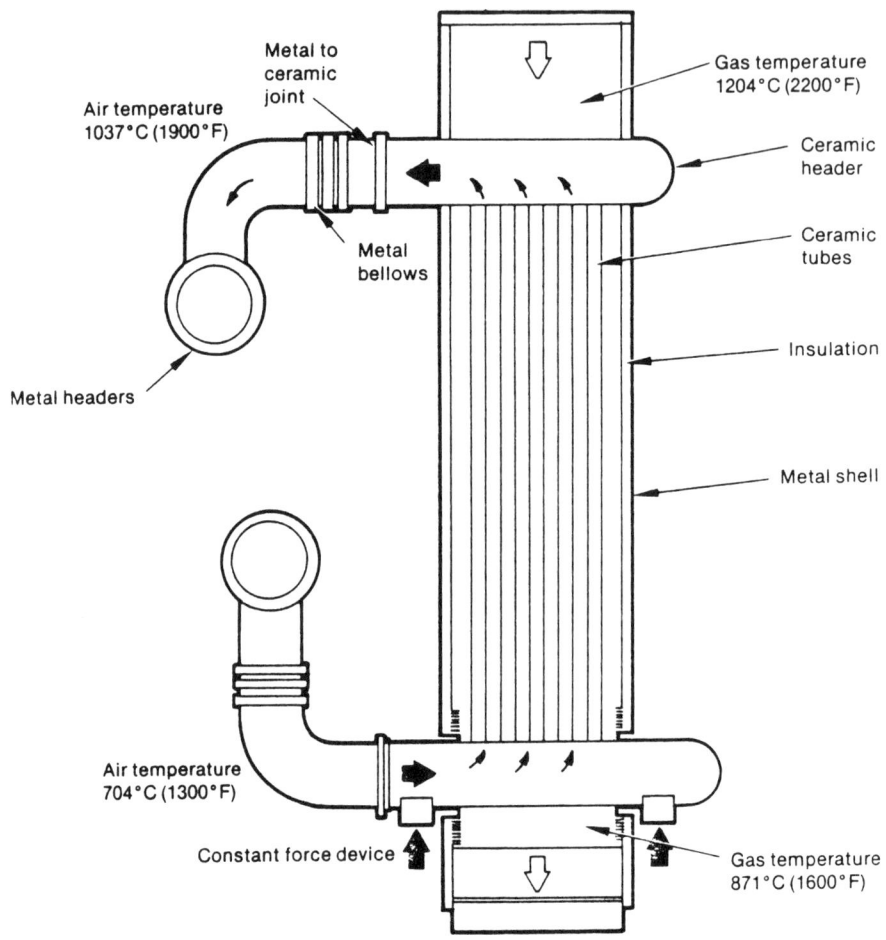

Figure 3.6 Cross-section of heat exchanger (Solar Turbines International).

Recuperator Designs and Applications 23

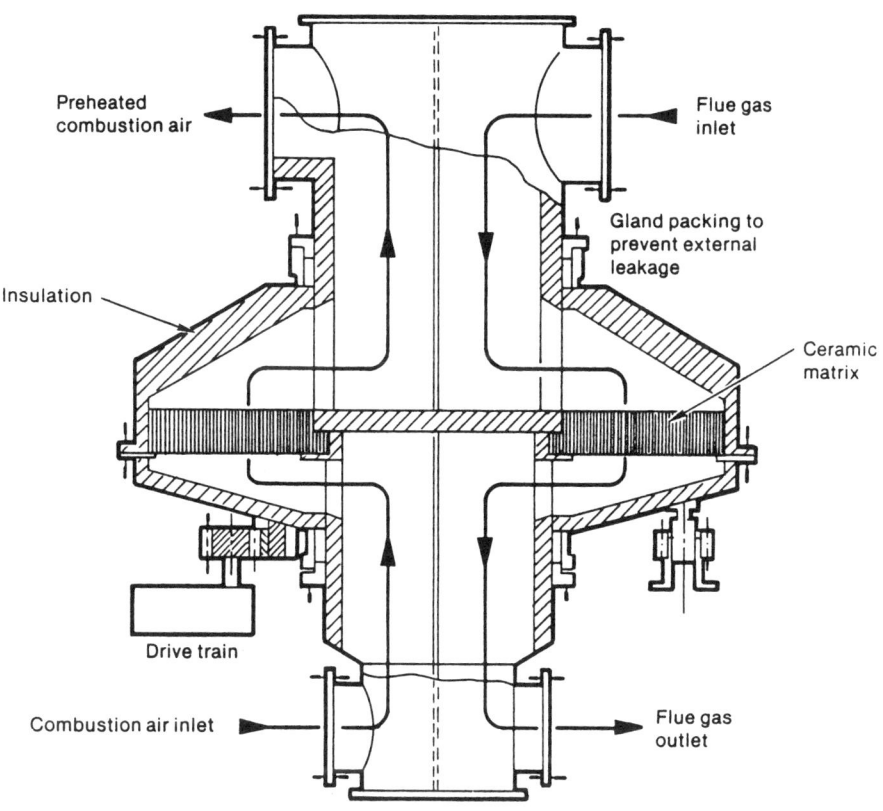

Figure 3.7 Schematic of the heat wheel.

glass remelting furnaces. The following subsections (3.2.1 thru 3.2.3) describe the process utilized to determine a number of potential concepts, the process of selecting which concepts to develop, and the detailed description of the three "selected" concepts.

3.2.1 Concept Development

A "new concept" is defined as one that is truly different from the ideas that are known to exist. In the case at hand, new concepts in the area of high-temperature recuperators were actively sought. The new ideas contained herein are a result of individuals' abilities to piece together seemingly unrelated processes and systems in new solutions to high-temperature recuperation. The ideas are not a collection of other people's high-temperature recuperation concepts. Upon development, some of the new concepts were found to resemble other people's ideas in this field. However, they are still included in this report since they may represent a slightly different approach to the problem.

The new ideas when first presented are in their infancy and are easy to dismiss as not being feasible. However, with the correct nurturing and encouragement these ideas may mature into workable systems that represent major breakthroughs in the area of high-temperature recuperation.

Since no limits were placed on the types of high-temperature recuperator ideas which were generated, a broad spectrum of the field was represented. To facilitate the presentation of ideas in this report, they have been separated into the following general groupings: material, fabrication, component, and process. Details and sketches of these ideas are presented in Appendices A, B, C, and D. Table 3.2 lists the new concepts and gives a brief description of the ideas. Reference should be made to the Appendices for a more complete presentation.

3.2.2 Concept Selection

After the concepts were initially developed it was necessary to select those concepts which would be developed further . Generally, a review of

TABLE 3.2 SUMMARY OF NEW CONCEPTS

New Concept	Brief Description
Materials (Ref. Appendix A)	
1. Crack Stopping Additives	Add small microspheres to ceramic stock to dissipate crack energy. Keep cracks below critical flaw size.
2. Piezoelectric Ceramics	Develop a ceramic that acts like a piezoelectric crystal. When strain occurs an electrical potential develops which attracts "plugging" particles.
3. Composite Ceramics	Build up a ceramic matrix with following features: cracks stop at interfaces, environment protective layer, acoustic emitters when overstressed, healing fibers.
4. Flexible Ceramic	Use composites with possible bendable fibers with gel-like matrix.
Fabrication (Ref. Appendix B)	
1. Woven Tube	Tubes are woven from a high temperature fiber into any shape (round, elliptical, square). Leakage is minimized or eliminated by tight weave or viscous liquid.
2. Wrapped Paper or Fabric	Fabric is woven with high temperature fiber. Tubes are constructed with multilayered, opposed bias wrapping. Viscous liquid to seal.
3. Ceramic Coated Fibers	Hollow ceramic fibers are made by coating volatile fiber with ceramic then firing. Tubes could be used in small Hx's or high temperature ultra filtration.
4. Compression Coating	Develop a skin that is capable of remaining in tension at temperature. Coat the ceramic with this material to keep it in compression.
5. Graded Material Tubes	Develop a fabrication technique to produce tubes with different properties on the ends than in the center. Tube middle is flexible, tube ends are rigid.
6. Wound Tube	Use rocket motor casing technology to wind high temperature fiber into tubes. Use viscous liquid as binder during the winding process.
7. Vibrating Fabrication	Use vibration during the firing and slip casting process to minimize voids and crack initiation sites.
8. Compressed Firing	Fire greenware in a compressed condition. Use very high pressures (1 million psi and up) to provide small transition pieces that will take tensile loadings.
9. Powder Production	Use a cavitation device--similar to the cavitation phenomena that destroys valve seats--to produce fine grained power for ceramic production.
10. Pliable Ceramic	Develop a pliable ceramic at operating temperatures that would relieve strains as they occur.

TABLE 3.2 (continued)

New Concept	Brief Description
Component (Ref. Appendix C)	
1. High-Temperature Paper Hx	Would be constructed from a thin ceramic paper. Could be assembled cheap enough so that it could be thrown away when fouled.
2. Inflatable/Expandable Tube Sheet	A standard shell and tube sheet arrangement would have some type of expandable media inside the tube sheet. At temperature this material would expand, seal the tubes, and put the tubes in compression.
3. Expandable-Extrudable Seal	The tube sheet in a standard tube and shell design contains a fluid under pressure that extrudes out past the tubes and forms a seal.
4. Stacked Cylinders with Viscous Liquid	Stacked flat disks make up a tube-to-tube sheet joint. Tolerance stackup allows for thermal growth and misalignment.
5. "Marching" Replacement Hx	A modular construction, expendable/limited lifetime component is proposed. Units would be installed at one end new and removed from the other end at end of lifetime.
6. Isostatic Joint Compression	Mechanical seals are loaded by means of an isostatic inflatable bag type of device. Bag is outside of high temperature region.
7. Liquid Covered Tubesheet	A viscous liquid forms a seal and damps vibrations in tube-to-tubesheet joint.
8. Alloy Migration	At temperature alloying elements migrate to cooler portions of tube. Hot tube-to-tube sheet becomes soft and pliable.
Process (Ref. Appendix D)	
1. Woven Heat Exchanger	A heat exchanger is completely woven from high temperature ceramic fibers. The fibers are used to weave the tube sheet as well as the tubes.
2. Paper Honeycomb Hx	High-temperature ceramic paper is used to assemble a honeycomb type heat exchanger. This Hx is assembled in modules.
3. Phase Change Shot Tower	A material that changes phase is allowed to melt in the hot exhaust gas stream, is delivered to the inlet air stream, and allowed to solidify.
4. Coated Sponge	A high-temperature sponge type material is used to help conduct heat away from flow passages that are present in the sponge.
5. Cross-flow Liquid Bath	A molten liquid is used to surround tubes in a heat exchanger configuration. Heat is transferred to the liquid by natural convection.
6. Tricycle Regenerator	A tube divided into three sections is rotated into and out of the hot gas stream. The section rotated out is cleaned.

TABLE 3.2 (continued)

New Concept	Brief Description
7. Gas Entrainment in Liquid	Exhaust gas is bubbled through a liquid heat transfer media. This media is then transferred to the air preheat side where air is bubbled through the hot liquid.
8. Oscillating Bubble	Bubbles are caused to extend into an exhaust stream. Pulsed bubbles cause air to flow in exhaust duct.
9. Lenses to Concentrate Heat	Use a fresnel or compound eye approach to concentrate heat. Concentrated heat source can heat incoming combustion air.
10. Ceramic Heat Pipe with spirally fluted tubing	Spiral flutes act as wick in a heat pipe configuration. Flutes eliminate the need for a separate wicking element.
11. Jet Impingement (with cyclone)	Exhaust gas is caused to jet impinge on a heat transfer surface. Heat is removed by combustion air from other side.
12. Tetrahedron Hx	A heat exchanger is constructed from easily formed tetrahedral shapes. These shapes are fit together to form heat transfer surfaces.
13. Membrane/O Molecular Sieve	A pelletized material absorbs N from cold preheat air and heats it. This material is transferred to hot exhaust where N is liberated. O is enriched in incoming combustion air. Could be used with raining bed type of technology.
14. Preheat Pelletized Feedstock	Small pelletized feedstock is heated in exhaust air stream.
15. Floating Solids Change Shape	Solids with aerodynamic shape float, heat up, and change shape. Hot shapes drop out of air stream are moved to preheat air, and cooled. Cooled shapes take aerodynamic shape again.
16. Chemical Reaction Pairs	Endothermic reaction in exhaust stream. Exothermic reaction in preheat air stream.
17. Cogeneration	Use thermionic stack walls to generate electricity. MHD, Brayton cycle, or density wheel.

Table 3.2 and Appendices A through D indicated that new materials, fabrication and component concepts are either already included in the new process concepts or could be used by most any of the process concepts to improve them in a later stage of development. Therefore, the concept selection was limited to the "process concepts."

Advantages and disadvantages were compiled for each of the process concepts. (It should be noted that some of the advantages and disadvantages associated with the selected concepts changed with further development; this is described later.) These are summarized in Appendix D. The 18 process concepts were then evaluated subjectively by using a "filtering system" (based on the evaluator's perception of the concept and his knowledge base in consultation with experts and literature sources). This filtering system is schematically depicted in Figure 3.8. Following is a description of the "filters."

3.2.2.1 <u>Basic Requirements Filter</u>. The basic requirements filter evaluated the process concept for ability of the materials of construction to operate at temperatures above 1900°F. It also evaluated the concept on how quickly it could be realized so that it would be ready for industrial development by the private sector in 5 to 10 years.

3.2.2.2 <u>Thermal-Hydraulics Filter</u>. The thermal hydraulics filter evaluated the process concept for its ability to exchange heat with reasonable pressure drops or flow resistance.

3.2.2.3 <u>Materials Filter</u>. The materials filter evaluated the process concept for material availability, material compatibility with the corrosive environments, and quantity of material necessary to construct it.

3.2.2.4 <u>Fabrication Filter</u>. The fabrication filter evaluated the process concept for the fabrication possibilities and limitations.

3.2.2.5 <u>Filtering Results</u>. The results of this filtering process are summarized in Table 3.3. Process concepts which passed through the filtering process are:

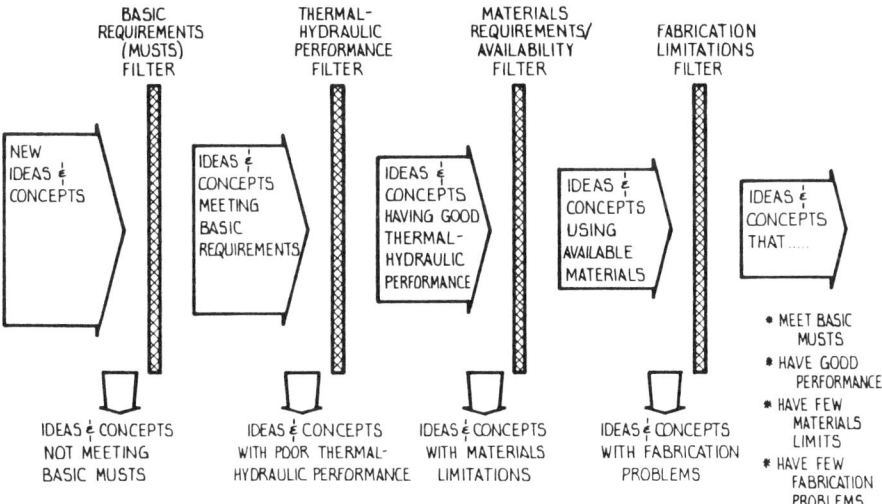

Figure 3.8 Preliminary filtering process for new ideas and concepts.

TABLE 3.3 SUMMARY OF PROCESS CONCEPT FILTERING EVALUATION

Idea	Basic Needs	Thermal Hydraulic	Materials	Fabrication
**1. Woven Tube/Tubesheet	Pass	Pass-Good Potential	Pass-fabrics Exist	Pass-Existing Technology
**2. Paper Honeycomb HX	Pass	Pass-Good Potential	Pass-Paper & Fabric	Pass-Technology Exists
**3. Phase Change Shot Tower	Pass	Pass-Good Potential	Pass-Questionable Material Being Ident.	Pass-Technology Exists
4. Coated Sponge	Pass	Questionable-High P Questionable	Pass-Fibers Available	Pass-Coating Technology Exists
5. Cross-flow Liquid Bath	Pass	Questionable Heat Nat. Conv. Coeffic. Lower HT	Questionable-Large Quantities	Pass-Existing Technology
*6. Tricycle Regenerator	Pass	Pass-Good Potential for Minimizing Effects of Dirty Exhaust Streams	Pass-Existing Materials	Pass-Existing Technology
7. Gas Entrainment in Liquid	Pass	Questionable-Large Pumping HP	Questionable-Questionable Material	Pass-Assume Buildable
8. Oscillating Bubble	Questionable-5-10 Years	Questionable-Unknown	Questionable-No Material Available	Questionable-Unknown
9. Lenses to Concentrate Heat	Pass	Questionable-Questionable Effectiveness	Questionable-Possible Materials Problem	Pass-Technology Exists
*10. Ceramic Heat Pipe with Spirally Fluted Tubing	Pass	Pass-Good Potential	Pass-Lab (LANL) Demo	Pass-Lab (LANL) Demo
11. Jet Impingement (w/Cyclone)	Pass	Questionable-High P & Fouling	Pass-Material Exists	Pass-Technology Exists
12. Tetrahedron HX	Pass	Questionable-High P & Dead Space	Pass-Material Exists	Pass-Technology Exists
13. Membrane O Molecular Sieve	Pass	Questionable-Oxygen Enrichment Contrib.	Questionable-Materials not Identified	Pass-Technology Exists
*14. Preheat Pelletized Feedstock	Pass	Pass-TECO Demo	Pass-But not all Processes	Pass-TECO Demo
15. Floating Solids Change Shape	Questionable-5-10 Years	Pass-Good Potential	Questionable-Unavailable Materials	Pass-Assume Buildable

TABLE 3.3 (continued)

Idea	Basic Needs	Thermal Hydraulic	Materials	Fabrication
16. Chemical Reaction Pairs	Pass	Questionable-Large Pumping HP	Questionable-Unidentified Compounds	Pass-Assume Buildable
17a. Cogeneration: Low Temp MHD	Questionable- 5-10 Years	Questionable- Unknown	Questionable-Unknown Material	Pass-With Development
17b. Cogeneration: Thermionics	Questionable- 5-10 Years	Questionable- Unknown Configuration	Questionable-Unknown Material	Pass-With Development

*Ideas passing through all of the filters.
**Ideas passing through filters and through programmatic constraints.

- Woven heat exchanger

- Paper honeycomb heat exchanger

- Phase-change shot tower

- Tricycle regenerator

- Ceramic heat pipe with spirally fluted tubing

- Preheat pelletized feedstock.

Note: This filtering evaluation was done subjectively such that it was based on the technology available (with a reasonable amount of research) to the personnel performing the selection. For example, the coated sponge concept passed the basic requirements filter because the evaluator was able to identify ceramic materials which can be used to fabricate the concept and which could withstand exposure to the high temperature corrosive streams. It did not pass the thermal-hydraulic filter because the evaluator thought that the pressure drop across the sponge would result in a great fan power cost penalty, a disadvantage thought not to to be counterbalanced by the advantages. It passed the materials and fabrication techniques filters because the evaluator thought that the ceramic sponge-like matrix could be made from ceramics analogously to foaming a plastic or a metal. This means that some of the other process concepts could potentially be developed if further information could be made available.

Of the process concepts which passed through this filtering system, the preheat pelletized feedstock concept was eliminated based on the inapplicability of the concept to the three selected process streams discussed previously. The tricycle regenerator process concept was eliminated because it can be classified as a periodic heat process being developed for gas turbine applications by a number of investigators including Ford, Midland-Ross, Solar Turbines, General Electric, and Pratt & Whitney. Although the fouling/cleaning cycle is an innovative adaptation

of the heat wheel, it was not felt that it would represent a large enough
technological advance to merit further development in this investigation.

The ceramic heat pipe process concept was also eliminated because
there are a number of investigators including Los Alamos National
Laboratory (LANL) who are already developing it. The LANL heat pipe
concept is considered later in the comparison/evaluation section.

The phase-change shot tower, woven heat exchanger, and the paper
honeycomb heat exchanger concepts were selected for further development and
final comparison and evaluation with emerging and state-of-the-art
technologies.

3.2.3 Detailed Description of EG&G Concepts

The following descriptions of the EG&G concepts were developed as part
of the conceptual design process. The level of detail is not consistent
nor should it be since the concepts were developed to different degrees.
The concepts are still in the developmental stage and many unknowns may
still be identified which may be further investigated if the basic concept
is evaluated to be feasible. These descriptions do not represent fully
developed conceptual designs, but discuss some of the potential advantages
and disadvantages for comparative purposes only. These concepts are sized
in Section 4.0 except for the phase-change shot tower which is sized in
this section.

3.2.3.1 Phase-Change Shot Tower. The phase-change shot tower heat
exchanger uses small droplets or beads of a working fluid to transfer heat
from the high-temperature exhaust gas stream to the low-temperature preheat
air stream. Solid beads are dropped into the top of a refractory-lined
tower. The exhaust gas flows upward through this section of the tower and
as the beads fall through the hot gas, they are heated and melt into
droplets of liquid. Meanwhile, the exhaust gas is cooled and exits the top
of the tower. Because some of the liquid droplets may coalesce, they are
allowed to fall and form a pool on top of a sieve distribution plate half
way down the tower. The liquid drips through the sieve and forms the

required sized droplets. These droplets then fall through the cool preheat air, which is flowing upward in this section heating the preheat air and resolidify into beads. At the bottom of the tower, the beads are collected and conveyed to the top of the tower where they repeat the cycle (see Figure 3.9).

An important aspect of the phase-change shot tower concept is the selection of the working fluid. The following requirements were established to aid in this selection:

- The fluid must be nonoxidizing

- The fluid must flow (like a liquid) between the temperatures of 200 and 2300°F for the soaking pit application, between 200 and 2000°F for the aluminum melting application, and between 200 and 2700°F for the glass melting application; it must not flow (like a solid) at lower temperatures.

Certain preferred qualities, as listed below, were also established to aid in the selection of a working fluid:

- The fluid should have a high heat capacity

- The fluid characteristics should not be adversely affected by gases or particulates in the waste stream

- The fluid should have a high latent heat value.

Once it was determined that various metals would not meet even the basic requirement of a working fluid for the short tower, glass was chosen even though it does not technically change phases but transitions at a point where it does not flow. It will not oxidize and its composition can be varied so that it flows in the desired temperature ranges. It has a high heat capacity and will absorb all particulates and gases in the exhaust stream. Glass can also be quite inexpensive.

Recuperator Designs and Applications 35

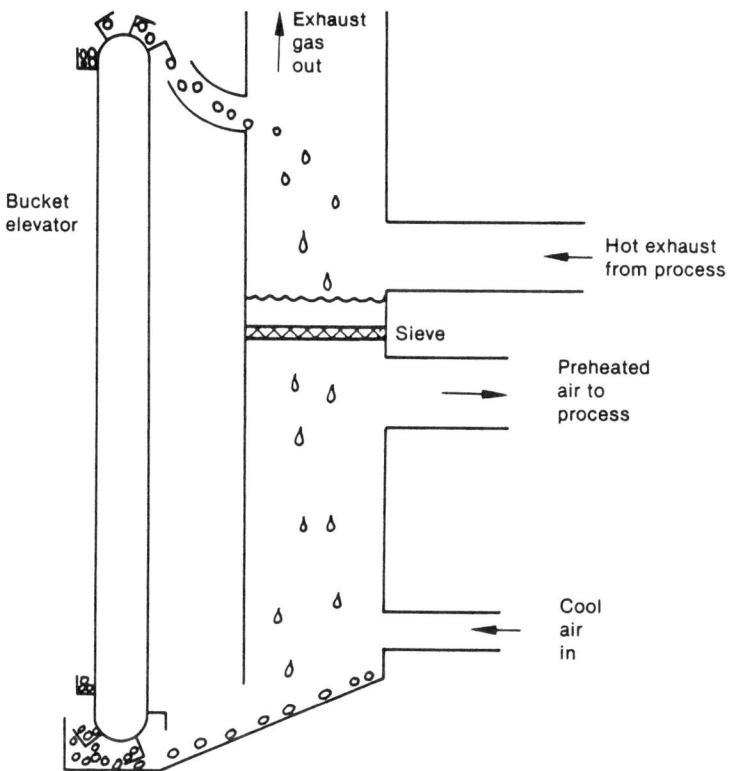

Figure 3.9 Phase-change shot tower.

Once the working fluid was selected, specific design features of the shot tower were developed. The calculations which were performed are included as Appendix E. Decker's "Falling-Bead Dry Cooling Tower" and Perry's Chemical Engineers' Handbook were used as references. A small enough bead size was used so that the heat transfer resistance between the bead surface and the gas became the controlling factor. The bead size was made as small as practicable so as to maximize heat transfer rates. The height of the heating section of the tower was calculated so that a 0.031 in. diameter bead would be heated through to its center in the time that it would take to fall through that section. The cooling section was assumed to be the same height as the heating section. The total height of the tower was determined to be 150 ft. The inside diameter of the tower was sized to accommodate the gas flow rate at its hottest temperature (and therefore its lowest density). The inside diameter was calculated to be 14.5 ft. The working fluid must transfer enough heat to the preheat air to heat it to 2000°F at high-fire conditions. This requires a heat transfer rate of 76,000 Btu/min. The working fluid flow rate required is 140 lbm/min.

Disadvantages--Although absorption by the glass of flue gas particulates may be a desirable feature, the properties of the glass will change considerably with the amount of particulate absorbed. A method of keeping the glass properties constant or within some specified working range must be found in order to use glass for long periods of time. One way to adjust glass properties is to use additives; however, this may be a costly alternative. Using the glass for one day or for even a shorter period of operation are other possible costly alternatives.

Glass is a very erosive material and will not lend itself well to the tower durability. Another possible problem is adhesion or solidification of the molten glass on the tower walls. It may be possible to install baffles or swirl air up through the tower. This would help the erosion problem and keep the glass from hitting the walls, but would contribute a large expense to the already exorbitant tower price.

Glass is not a very good conductor of heat, another reason the beads are very small. A material which met the material requirements and conducted heat well would be more desirable. Glass also does not have a latent heat of fusion; therefore, the design cannot take advantage of both the latent heat of fusion and the sensible heat.

Advantages--The phase-change shot tower concept has the advantage that the technology exists for lower temperature processes, where the towers are made shorter by the use of baffles thereby allowing a smaller unit capable of the same performance as the described tall units. The particle size is very closely related to the tower height since the required bead drop height increases as the bead radius increases. The glass beads were made as small as possible to reduce the height of the tower; however, to obtain such tiny particles, the holes in the sieve must be even smaller than the bead. This leads to such questions as:

- Will the material even flow through such small holes

- Will the holes be blocked by the particulates collected in the fluid

- Will the sieve need to be replaced often because of the corrosive nature of the working fluid

- Will the sieve need a cutter on it to produce the correct bead size

- Will the working fluid flow through the holes fast enough to maintain the design flow rate?

These are all valid questions which require resolution before implementing the design. They are highly dependent on the properties of the sieve material chosen and upon the characteristics of the working fluid: i.e., viscosity, fluidity, erosive nature. If the sieve is continually blocked, erodes quickly, or does not allow the fluid to drip through, it will become a major expense and the weak point in the design.

Conclusions--Based on the working fluid which was chosen as a result of this developmental work, this concept would require a much larger space than the design requirements will allow (150 ft height compared to 4 ft allowed). Therefore, it will not be further evaluated in this report. Control of the residence time of the beads by varying the airflow might lend to the feasibility of this concept.

3.2.3.2 Ceramic Fiber Heat Exchanger Concepts. The other two selected EG&G concepts are similar in the materials chosen for their construction: ceramic fibers, cloths, or papers. (There are a number of these types of ceramic materials currently commercially available. Some of these are listed in Table 3.4.) The EG&G tube-in-shell design would consist mainly of ceramic fibers woven to obtain the desired configuration. The EG&G finned plate design would be constructed of prefabricated sheets of ceramic paper or cloth. Any of the current state-of-the-art, emerging, or new designs could also use any combination of these materials. But, the EG&G concepts have also incorporated unique joining, fabrication, and configuration features which are partially made possible by the selection and use of these materials. The two EG&G concepts are described in more detail in the following sections.

Woven Tube Ceramic Heat Exchanger--The woven tube ceramic heat exchanger concept is a basic tube-in-shell design, differing from the traditional design in the materials fabrication methods used. In this concept, the material used is a fiber woven from high-temperature ceramic material into a one-piece tension structure composed of tubes and perforated tube sheets with walls 0.060 in. thick or less. (See Figures 3.10 and 3.11.)

Weaving is already an automated, well-established industry and only relatively minor changes may be needed to handle ceramic fiber, mainly related to fiber flexibility. The computer and computer-controlled looms make possible a great variety of design variations. Tubes can be of any shape or configuration determined to be effective for a specific waste stream. Tubes could be arranged as desired for fouling and corrosion resistance, ease of cleaning and inspection, and stress field optimization

TABLE 3.4 SOME COMMERCIALLY AVAILABLE CERAMIC FIBERS, CLOTHS, AND PAPERS APPROPRIATE FOR EG&G CONCEPTS

Ceramic Material	Available Form
Silicon Carbide	Nicalon^R SiC continuous fiber[a] (500 filament)
	Nicalon^R SiC woven cloth;[a] (warp: 32 y/in., weft: 32 y/in.)
	Nicalon^R SiC paper[a]
Alumina	Fiber FP^R yarn[b] (210 filament/yarn)
	Fiber FP^R satin cloth[b]
	Alumina paper (0.040 thick w 5-6% organic binder)
Zirconium Oxide (stabilized)	Zirconia felt (0.050 thick)
	Zirconia cloth, tricot knit (0.015 thick)
	Zirconia cloth, satin weave (0.030 thick)

a. Nicalon is a Nippon Carbon Co. Trademark.
b. FP is a DuPont Trademark.

40 Ceramic Heat Exchanger Concepts and Materials Technology

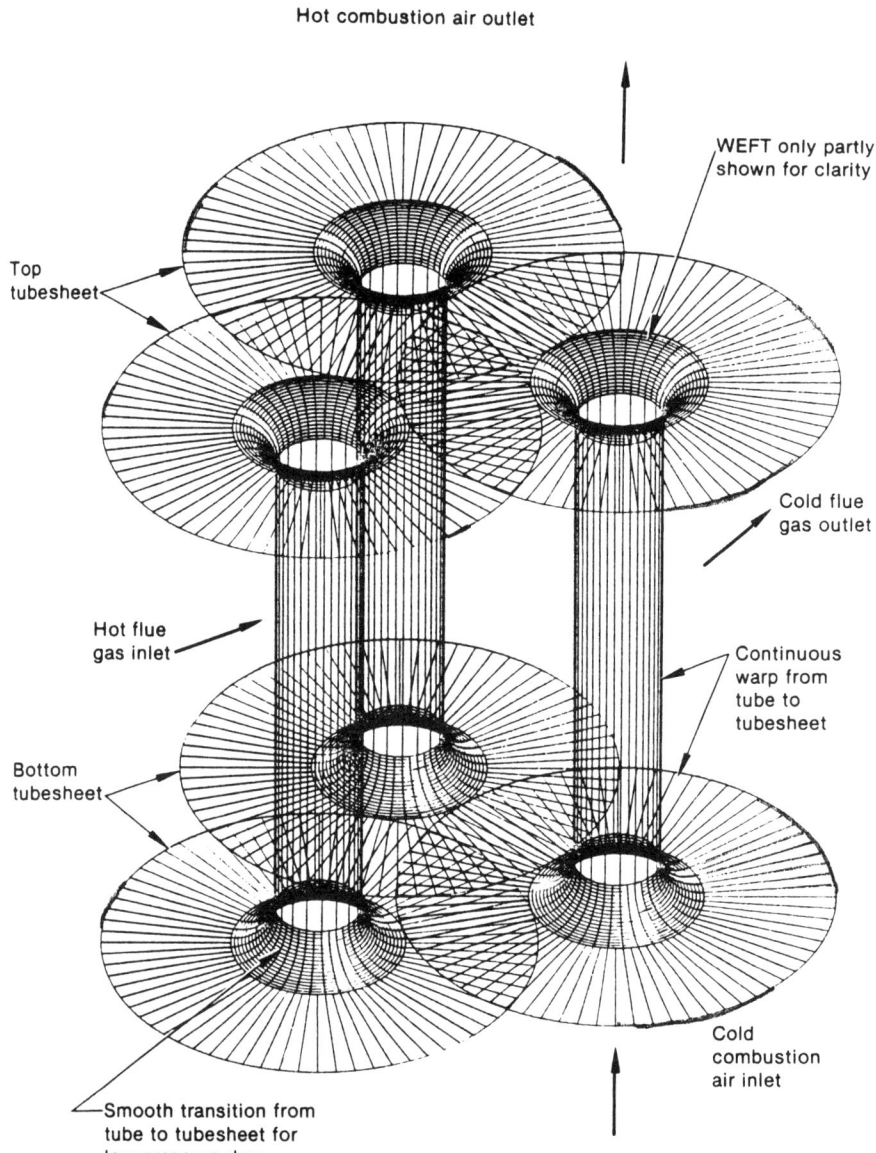

Figure 3.10 Woven tube heat exchanger isometric.

Recuperator Designs and Applications 41

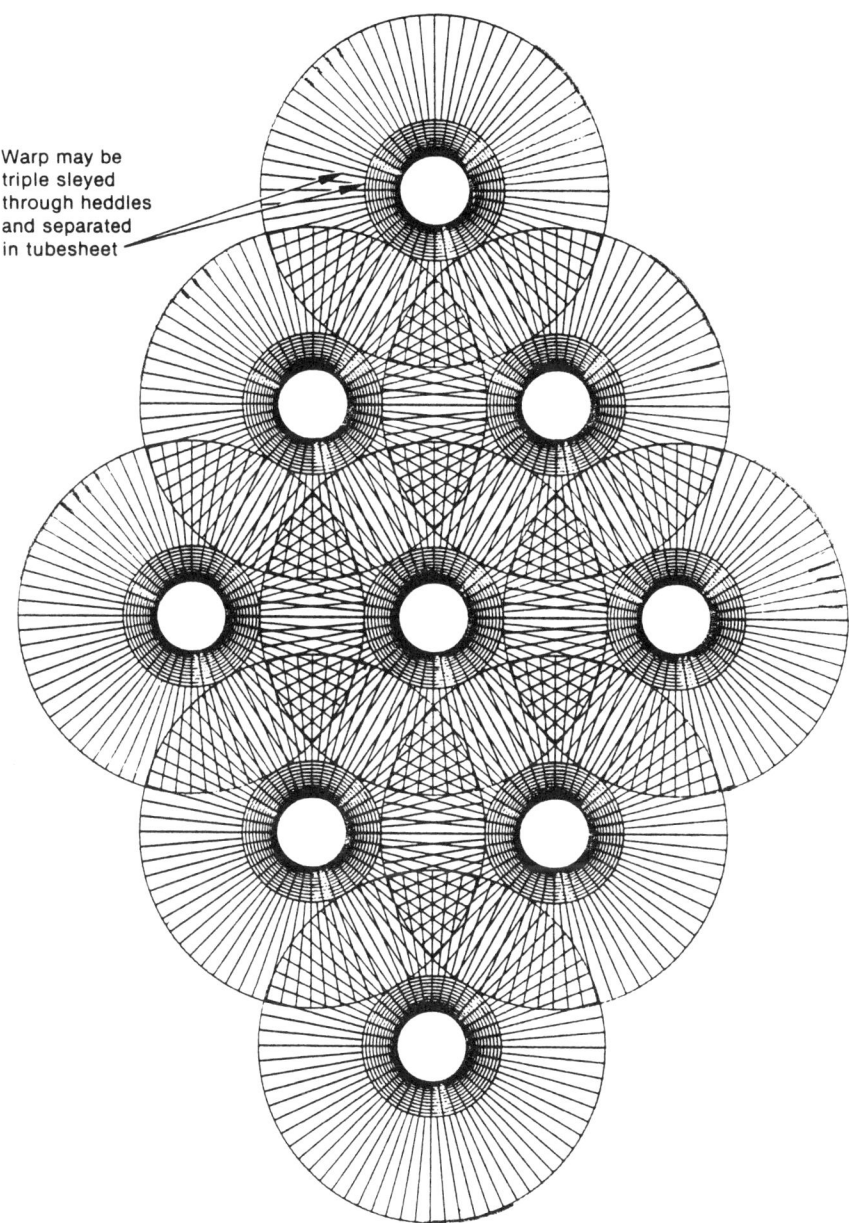

Figure 3.11 Woven tube heat exchanger tube sheet end view.

or redirection. The computer could optimize a heat exchanger design for each application if a detailed analysis model is developed and if the looms are also computer controlled.

If 3-D weaving is too undeveloped to obtain the desired shapes at one time, an alternate construction method may be used in cases where heat exchangers could be "built up" from three parts: a flat woven tube sheet with holes woven or cut, a woven/braided tube, and a transition piece to connect the two. If weaving on a 2D loom is too difficult, the parts could be sprayed and molded like a fiberglass boat or wound like a rocket casing. Rather than weave the heat exchanger as a unit, modules of the heat exchanger may be woven and then joined together.

Advantages--The one-piece (unitized) construction has the potential to eliminate leaks between tubes and tube sheets, and the uniform material thickness has the potential to eliminate thermal gradient stress caused by material thickness changes. The radiused transition from tube to tube sheet may allow the smooth transfer of loads, from pressure or thermal expansion, without stress concentration such that stress fields are more constant and change more gradually when they do change; and if a part is overstressed, some fibers break, but others carry the load and the recuperator does not break, as would a unit made of monolithic ceramic material.

Another advantage of the woven heat exchanger concept is that the thin and relatively lightweight fabric has only the required amount of material which helps to minimized costs . This can be a definite advantage if fabrication costs are not out of proportion.

Since ceramics display excellent compressive strength and fibers have excellent tensile strength, a ceramic matrix could withstand compression and the fibers would withstand tension. With a module design, a module could be more easily replaced in case of tube damage or fouling.

If a ceramic fiber matrix is attacked by a particular gas stream constituent, the matrix could be coated with a resistant material or another compound could be added to the matrix. This way, a matrix or coating could be compounded to exhibit specific properties, or stratified for special chemical resistances of sacrificial layers in the manner of cathodic coatings.

The concept lends itself to the techniques of NDE and acoustic emission, especially since the diameter of ceramic fibers in a heat exchanger can be different. If tracer strands, compounded to break well before the main fibers, were included in the weave, acoustic emission detectors could listen for fracturing fibers and identify set point stress levels and their location. This information could be used to predict maximum safe stress versus operating condition variables, stress field concentrations, and to indicate impending failure. This technique could also be used to indicate when fatigue limits are imminent and replacement is required.

Varying fiber diameters instead of materials can be useful in the heat transfer area. The ratio of surface roughness to hydraulic diameter for a specific operating condition can be varied to yield the desired heat transfer coefficient and pressure drop. Braiding a large diameter fiber spirally into a tube will act as a tubulator, which is much less parasitic to gas flow than the twisted tape kind. Adding or making fins integral to the tube structure is possible, but adds complexity and cost. The idea of using wooly yarn strands to act as pin fins is much simpler. On the gas side of the heat exchanger, the wooly fibers could induce turbulence in addition to acting as pin fins. Properly positioned wooly fibers could delay boundary layer separation and may damp vortex shedding oscillations. However, too much "fur" may not increase the heat transfer surface area or improve the flow conditions and fluid properties but may only increase the resistance to conductive heat transfer and thus increase the overall resistance to heat transfer.

In long tubes, weft threads would take the hoop stress and could have a smaller cross section than warp threads, which run longitudinally. Warp threads must bear tube weight and transfer loads to the tube sheet. At the transition area between tubes and tube sheet, weft threads could be increased in size to counter the effects of supporting the warp threads at the 90-degree bend. With the woven tube concept, computer-controlled looms can make almost any smooth transition geometry.

Disadvantages or Unknowns--There are potential problems or unknowns with this concept as with any new idea. Chemical attack by gas stream components is one unknown. This might be solved by coating the fibers with an impervious material. Another problem is compressive strength. Embedding the ceramic fibers in a rigid matrix could increase buckling of the unsupported fibers but also could prevent environmental attack.

Leakage through the fabric may be a problem because no matter how tightly a fabric is woven, some leakage will occur; however, this may be negligible. The answer to the problem may be in the material itself--SiC forms an oxide coating on its outer layer at high temperatures, and zirconia fibers sinter to each other above 2500°F. These coating/binding mechanisms may seal the pores. Other possible solutions may be a viscous glass in the press, or a matrix encapsulating the fibers.

Flow-induced vibration may be another problem caused by woven tube construction even though woven ceramic fabric is more rigid than a textile material. A viscous glass coating would dampen oscillations, and a rigid matrix would also stiffen the structure.

A major problem for an uncoated tension structure may be tube sheet to duct transitions and ensuring that, at high temperatures, the fabric is still loaded in tension. For a rigid matrix-fiber structure, the transition area is not as much a problem because, in addition to clamping, bonding with refractory grout or cement is an option to deal with compression and shear loads. Extending the tube sheet weave, or tensile fibers, past a seal and through an insulating zone to a cooler clamping area may be the answer.

Other unknowns are high temperature strength, fatigue, creep, and the effects of various gas stream constituents on woven ceramic structures. In most cases, however, these variables are also unknown for nonfibrous ceramics.

Ceramic Paper Heat Exchanger--The ceramic paper heat exchanger concept is a plate and fin design made from a thin, 0.060-in. ceramic paper material. The term "paper," as used in the ceramic paper heat exchanger concept, refers to any kind of ceramic fiber product including any flat, reinforceable ceramic fiber product such as ceramic cloth or crossed tape layers. If stress fields are adequately defined and their directions known, oriented-fiber products such as cloth or tape may be used successfully for this application.

In this concept, flow passages are honeycomb-shaped ceramic paper corrugations sandwiched between plates of the same material. The parts are rigidized and bonded together. The heat exchanger is built up of successive layers of these sandwiches, stacked with the corrugations, or spacers, either parallel to the layer beneath (counterflow configuration) or at 90° to the layer beneath (cross-flow configuration). Figures 3.12 through 3.18 show the cross-flow configuration and Figures 3.19 through 3.21 show the counterflow.

The cross-flow configuration of the ceramic paper heat exchanger concept uses entrance ramps to divert and accelerate the waste stream flow into the heat exchanger core. The entrance ramps are bonded to and feed loads to the corner posts. The intersection of the entrance ramps of each level forms a locator for the posts and a seal for each level. Corner posts are grouted to the ductwork in a single-unit installation. In a multimodular unit, corner posts are sealed to their neighbors.

In the cross-flow configuration shown in Figure 3.20, four corner posts serve as entrance/exit ramp bond areas for each level and define the leading edge limit for each ramp. The honeycomb corrugations serve as plate spacers and fins and determine the flow direction for each level.

With some slight modifications to the corrugations to provide better flow direction and reduced flow resistance, the counterflow configuration is the diagonally stretched version of the cross-flow configuration. Flow travels in a very long, very narrow X fashion, the length-to-width aspect ratio determining the exact shape of the X.

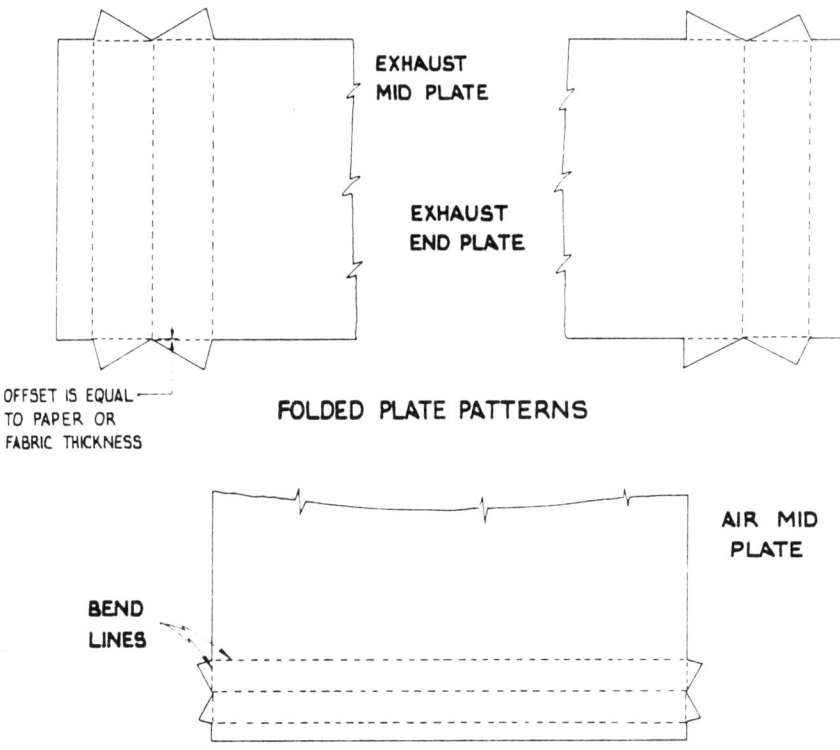

Figure 3.12 Folded plate patterns for paper finned plate design.

Recuperator Designs and Applications 47

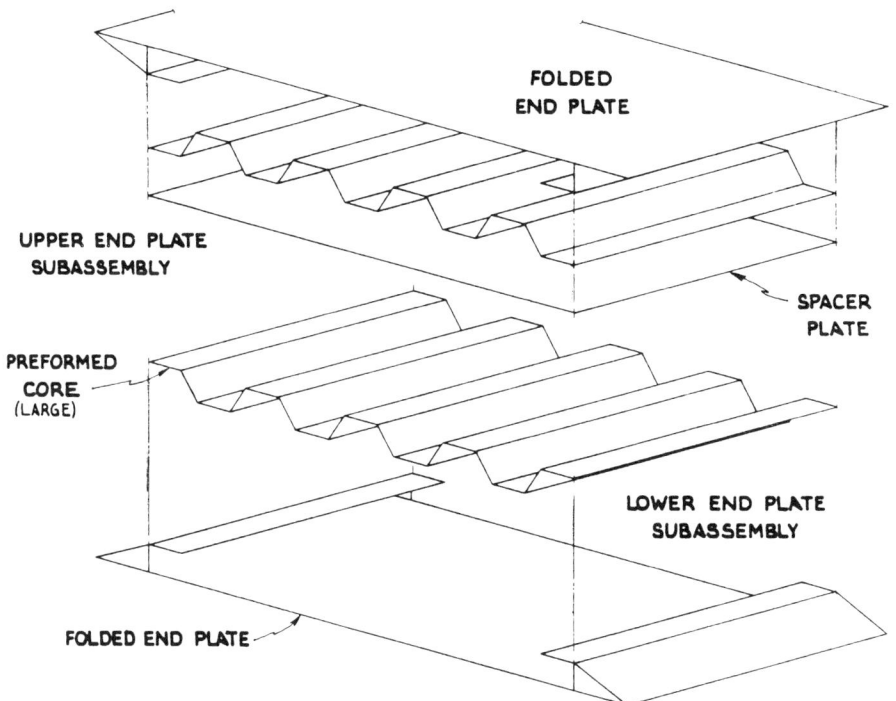

Figure 3.13 End plate subassembly for paper finned plate design.

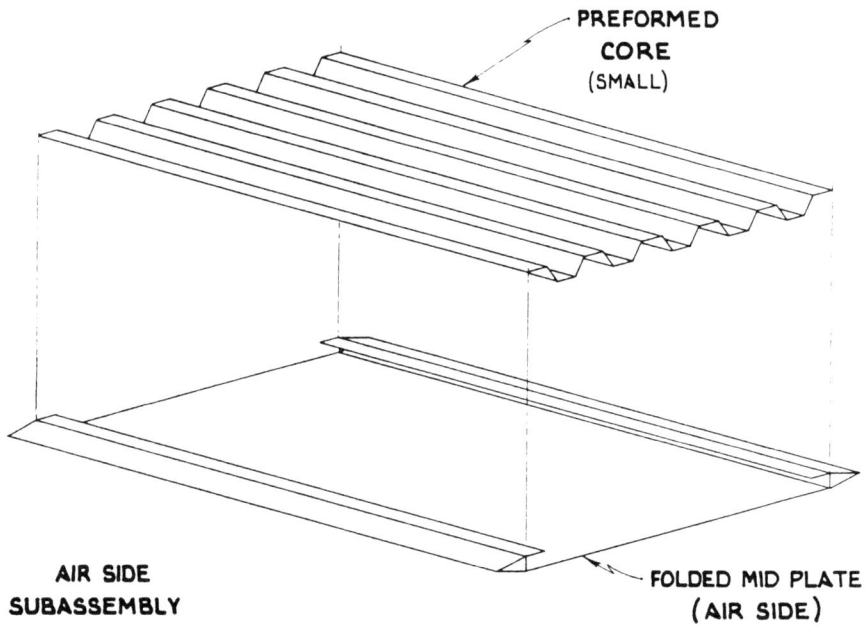

Figure 3.14 Air side subassembly for paper finned plate design.

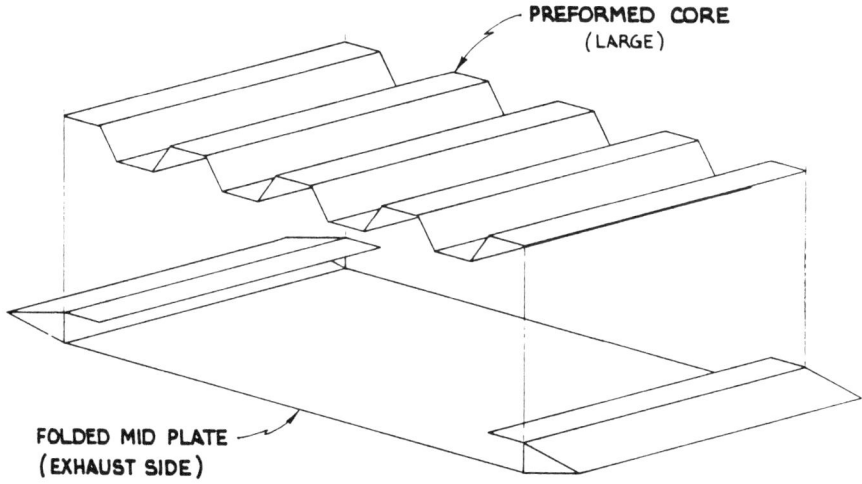

Figure 3.15 Exhaust side subassembly for paper finned plate design.

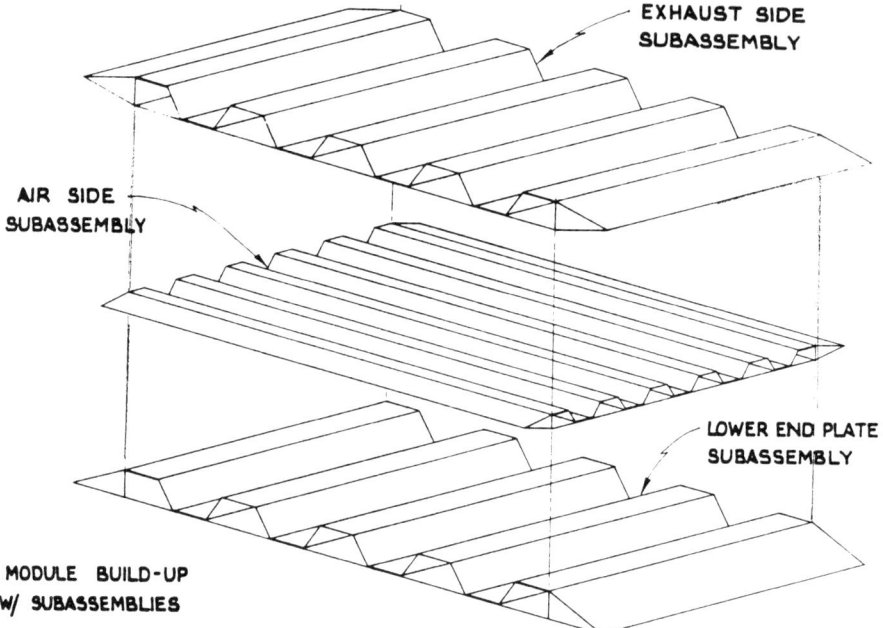

Figure 3.16 Paper finned plate module build-up with subassemblies.

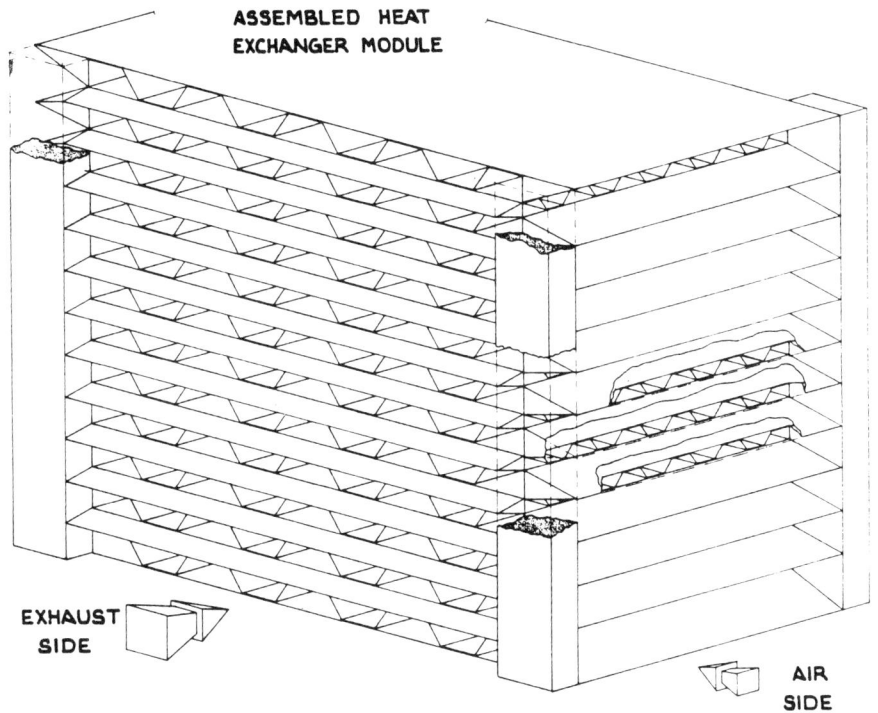

Figure 3.17 Assembled paper finned plate module.

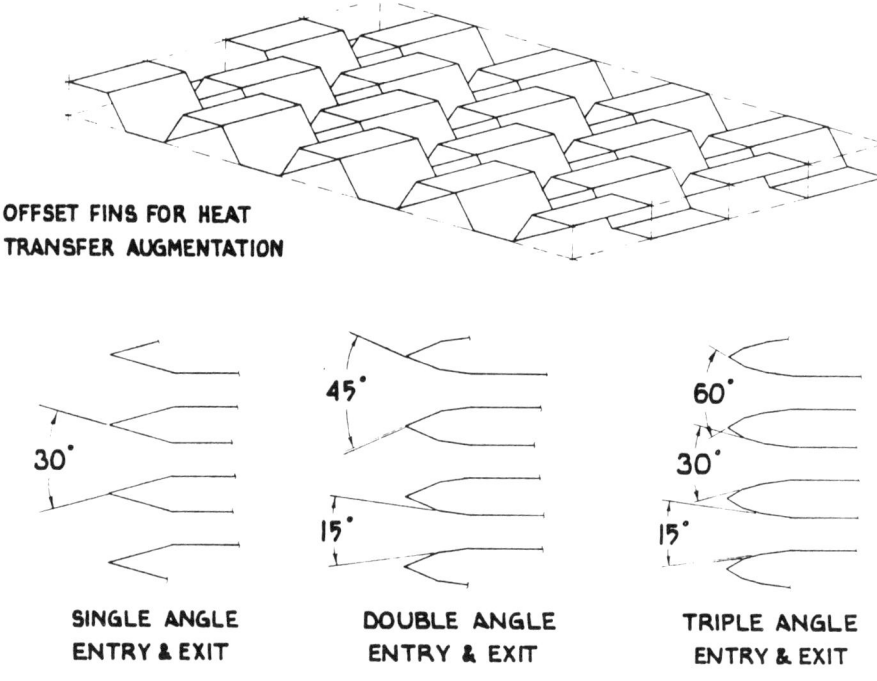

Figure 3.18 Ideas for increasing effectiveness of paper finned plate design.

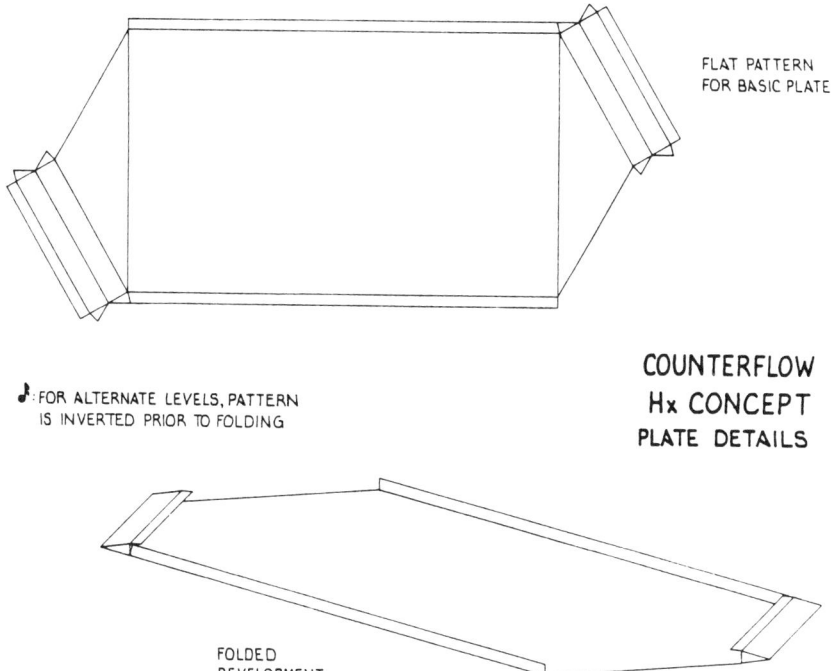

Figure 3.19 Flat layout and folded plate development for counterflow paper finned plate design.

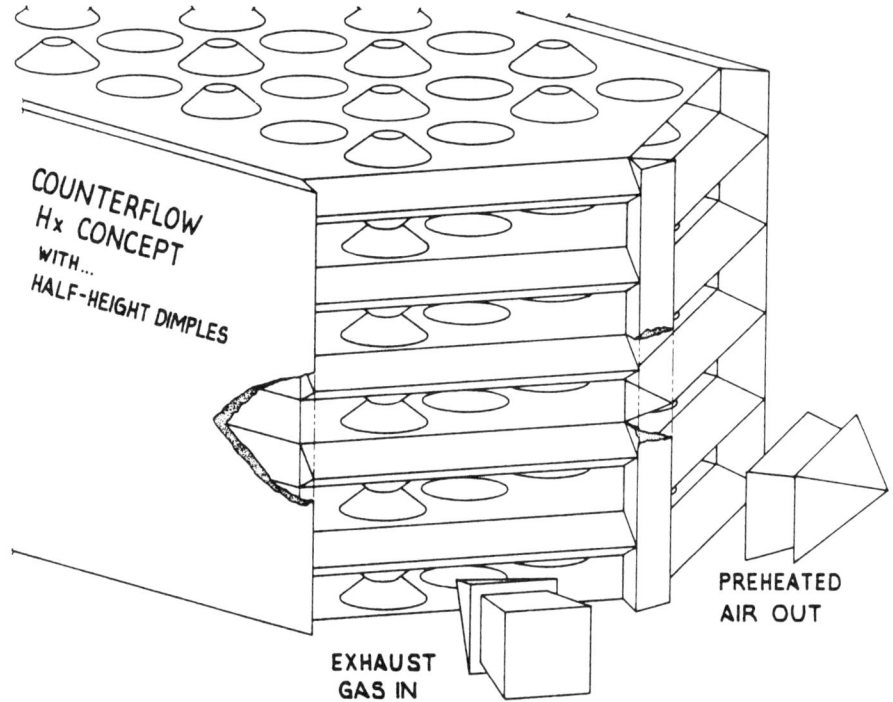

Figure 3.20 Dimpled stacked core of counterflow paper finned plate design.

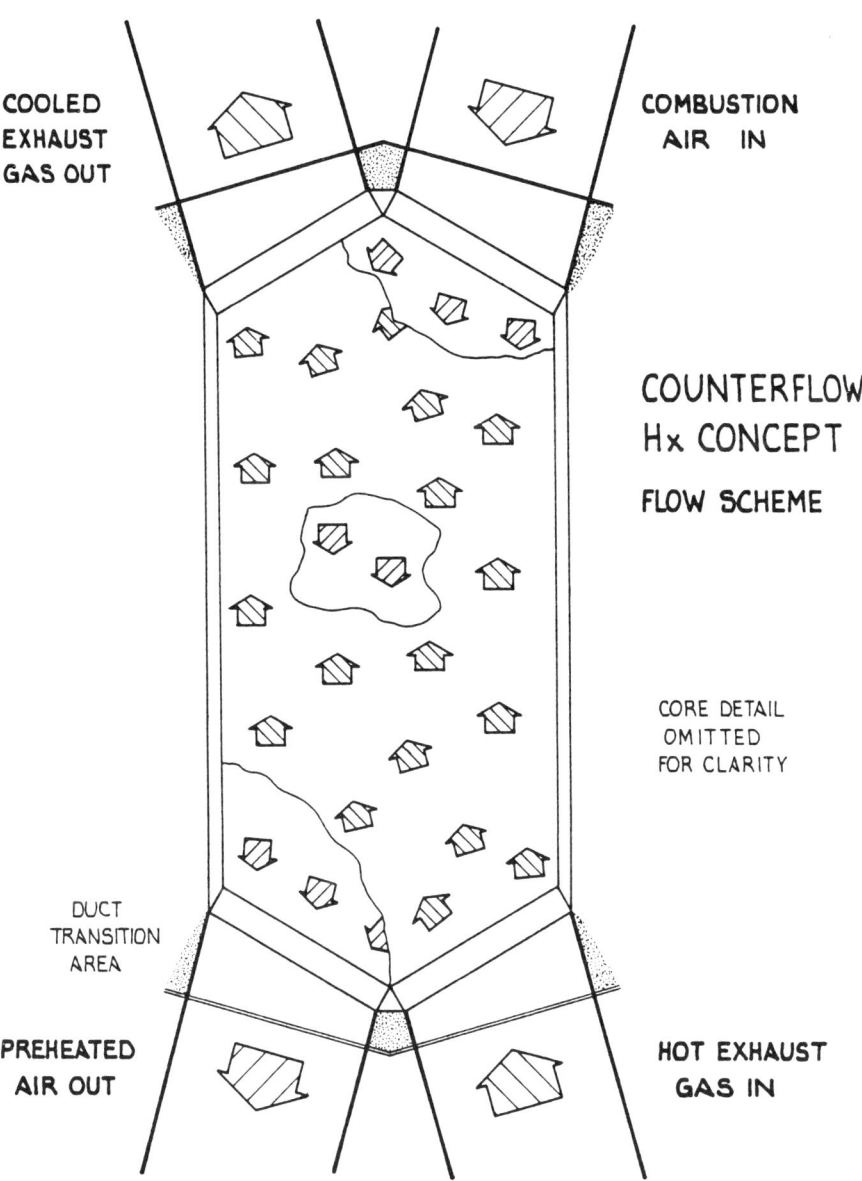

Figure 3.21 Flow arrangement of counterflow paper finned plate design.

Chemically resistant fibers may be needed for passages cooling exhaust stream gas, whereas less costly nonfibers could be used for the preheat air side of the core. Laminations with a chemically resistant side may be used, like the one-side galvanized steel used for automobile stampings.

In varying heat transfer, use can be made of variable weave pattern or tape fiber spacing to create periodic surface roughness. The heat transfer merits of wooly surface fiber ends have already been discussed in the woven tube concept description.

Concerning fabrication of ceramic paper heat exchangers, the cross-flow configuration can be produced in small modules, allowing a built-up core, or as a larger unit, eliminating module-to-module joints. The counterflow configuration is best restricted to a single unit core to eliminate potentially severe sealing problems. No matter which configuration is being fabricated, a maximum of seven different types of parts are needed; only four in some cases. Automated fabrication should present no problem; manufacturing equipment flexibility to handle a wide range of sizes is assured with computer control. Flat stock can be cut with laser or water jet cutters, and the basic instruction set would be common; locational coordinates would be the only change for each required size.

Advantages--The ceramic paper heat exchanger concept uses less materials than the traditional monolithic ceramic plate and fin heat exchanger and has the potential to be lighter. This lighter weight requires less support structure, making overhead or elevated locations more attractive.

In the paper concept, the inlet and outlet ramps (see Figure 3.20) should reduce some entrance and exit losses by gradually accelerating and decelerating the gas streams. The conceptual ramp design should also eliminate dead space. The ramp also provides more heat transfer surface area since the alternate stream between the open flow passages flows behind the ramp. In addition, the material temperature of the leading edges is always lower than the gas stream temperature, thus lengthening material life.

The thinness of the ceramic paper contributes to thermal shock resistance which lengthens material life. The paper also conducts heat more readily than thick materials although this effect may be negligible since convective heat transfer is the rate determining step. It is also more flexible than thick materials. In a ceramic material, this flexibility is an advantage since stresses cause material strain resulting in deformation instead of cracking.

The rigidizing compound applied to the ceramic paper could keep the paper subassemblies stiff enough to hold their shape during fabrication and provides a good bonding base. And when several stacked levels are bonded, the transfer of loads along multiple paths could reduce the loads at the joints and lower the overall stress levels. Redundant load paths offer another advantage in that if a fiber is overstressed and fails, its neighbors share the load and the part does not fail catastrophically. In a monolithic ceramic part, a small crack can grow rapidly and shatter the part.

Considering the rigidizer as the matrix of a composite, the ceramic fibers are encapsulated by the matrix, increasing their resistance to buckling, compressive loads, and tensile loads. Because of the random fiber orientations of a paper, fibers may experience pure tension, pure shear, pure bending, and any combination at the same time. The matrix supports the fibers and not only acts as a fiber-to-fiber compressive link but also reduces porosity, thus minimizing leakage and preventing corrosive waste stream components from attacking the thin ceramic material.

Disadvantages or Unknowns--Possible problem areas with this concept are leakage, fouling, and loss of strength at high temperature. Thin ceramic paper leaks more than thicker materials, but a rigidizing matrix could reduce leakage substantially. Also, mechanical abrasion and chemical erosion can weaken the thin material and cause leakage.

Poor fouling resistance is a serious problem with the entire range of plate and fin heat exchangers because of their flow channel size. In addition, heat transfer augmentation techniques such as offset strip fins, increased fins per inch, and small plate spacings add to the fouling problem. Fouling products can chemically attack the heat exchanger surface, but this can be at least partially prevented by a rigidizer matrix. The design parameters could be varied to minimize the fouling tendencies of any plate-fin configuration.

The effects of high temperatures are another potential problem. No test data are available on ceramic paper's high temperature strength, fatigue, and creep. However, it is known that material strength is definitely reduced at temperatures 80 to 90% of its melting point; to what extent is unknown.

3.2.4 <u>Detailed Description of Ceramic Heat Pipe Concept</u>

Heat pipes are evacuated tubes filled with a liquid called the working fluid. (See Figure 3.22.) Heat is applied to the evaporator end. Vapor is generated from the hot liquid and flows up the center of the tube to the other end of the tube (the condenser) where it condenses. The condensed liquid runs back down the walls of the pipe (gravity return) or, if the heat pipe has a porous wick next to the tube wall, runs back in the spaces between the wick and wall (capillary return). Heat pipes with wicks will work in any orientation with, without, or against the force of gravity.

Heat pipes conduct heat at very fast rates within certain limits depending on the temperature. The wicking limit occurs when the fluid at the condenser end cannot flow fast enough to the evaporator end through the wick. The wick then dries out in the evaporator region and "burnout" occurs. The sonic limit occurs when the vapor velocity goes sonic and the flow is choked. The entrainment limit is reached when droplets of liquid are sucked out of the wick into the vapor stream and not enough fluid can reach the evaporator end. The heat flux limit occurs at the evaporator end when boiling occurs and the generated vapor displaces fluid in the wick. (See Figure 3.23.)

LANL has had the best success with SiC heat pipes coated inside with tungsten by the chemical vapor deposition (CVD) method. A porous wick

Recuperator Designs and Applications 59

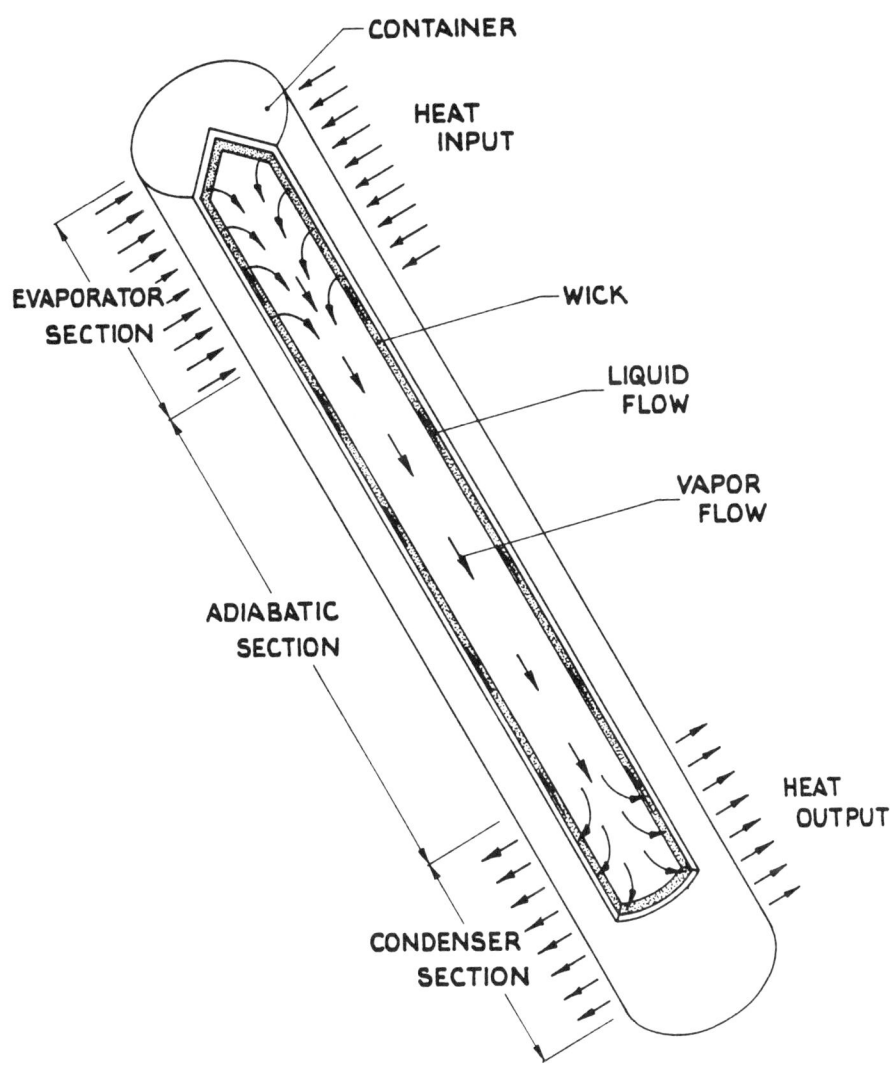

Figure 3.22 General arrangement of heat pipe.

60 Ceramic Heat Exchanger Concepts and Materials Technology

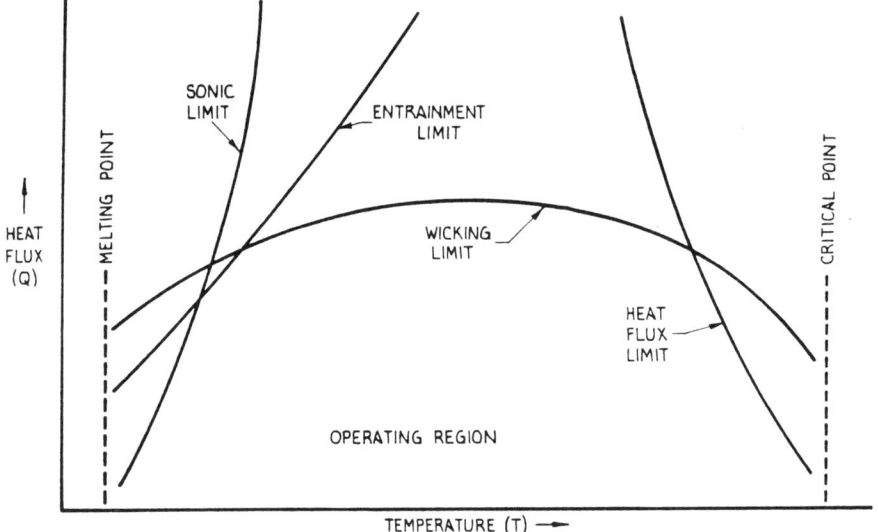

Figure 3.23 Limit relationships for heat pipe.

structure is achieved by allowing large grain growth to occur in the last part of the CVD process. One of the alkali metals, such as lithium or sodium, is used as the working fluid. A small tab of hafnium is required inside the tube to absorb any free oxygen molecules.

LANL's first successful high temperature ceramic heat pipe used CVD silicon carbide for a container material with a CVD tungsten coating inside and sodium as the working fluid. LANL operated these heat pipes in air with combustion gases at temperatures to about 1700°F. Peak heat transfer through the 0.75 in. diameter heat pipe was about 2 kW. Estimates of maximum operating temperatures are up to 2240°F with current technology.

Advantages--The heat pipe recuperator offers a major advantage in that if a pipe fails, it will normally not affect the operation of the other heat pipes and can be easily replaced.

Disadvantages and Unknowns--LANL has tried to develop various ceramic-to-ceramic and ceramic-to-metal bonding techniques for use as the hermetic vacuum seal. Difficulties encountered in developing these techniques forced the use of a tungsten-tungsten braze using palladium-cobalt brazing alloy. Development continues on welded end closures and also on an all ceramic glass seal.

One severe problem (other than end closure sealing) is the reaction of the working fluid to the inside tube wall. Sodium attacks silicon carbide readily, hence the necessity for the tungsten coating which reacts much slower with the sodium. LANL estimated a 20-year lifetime for a tube with a 0.010-in. thick layer of tungsten. They tried a heat pipe using zinc as a working fluid with a siliconized silicon carbide tube featuring a glass-sealed end closure. Although operated at 1520°F successfully, subsequent sectioning and examination revealed a reaction between the wall and the working fluid. LANL is continuing their materials compatibility studies.

Currently, sodium and lithium are the most frequently used working fluids. Unfortunately, they have a high reactivity rate with just about everything. Experts feel that group IIB metals have a lot of potential because the predicted reactivity rates with the heat pipe walls are much less.

Improvements and Their Advantages--Several ideas have been proposed to improve the performance/durability of the LANL heat pipe including varying the gas loading, tube materials, tube geometry, and heat exchanger configurations. Gas loading with noncondensible gas in variable volumes improves the performance of the heat pipe by lessening the thermal shock/strain on startup. It can also be used to vary the conductance to match high-fire/low-fire conditions. On-line monitoring of gas pressure could be used to indicate leak(s) in the heat pipes. The variable volume gas cylinders can be located outside of the flue stack in a relatively cool area where sealing is not so much of a problem.

Multilayer fiber/matrix composites can be developed so the interior matrix is working fluid resistant or, even better, inert. The outer layer of material forming the exterior of the heat pipe should also be minimally reactive with the waste gas streams thus ensuring a long lifetime for the tubes.

The convective heat transfer (the limiting factor in the rate of overall heat transfer) may be augmented by altering the geometry of the heat pipe. Noncircular cross sections such as elliptical or airfoil-shaped sections incorporating heat transfer augmentation techniques could be used to increase film coefficients. At the same time, the pressure drop would not increase as much as it would for normal circular tubes, thus improving the overall performance of the heat pipes. These geometries do not have to be constant throughout the heat pipe. They can vary in order to meet specific requirements, like fouling resistance, minimal pressure drop, and enhanced heat transfer in the gas stream zone, and interface and insertion allowances in the partition/seal zone.

LANL and Garrett have envisioned a counterflow arrangement of gas/air streams with the heat pipes inserted in a cross-flow manner to transfer heat from one side to another.[15] A modification to improve this basic configuration would be to vary materials and working fluids depending on position (according to temperature) in the heat exchanger. Further modifications to increase the heat transfer area while minimizing the volume could be made.

References

1. J. J. Cleveland, J. M. Gonzalez, K. H. Kohnken, "Ceramic Heat Recuperators for Industrial Heat Recovery;" DOE/CS-40174-T2 (GTE Products Corp.) 1980.

2. R. Dorazio, J Ferri, W. Fulerson, J. Gonzales, K. Kohnken, W. Mahoney, "Technology Acceleration Program for the GTE Ceramic High-Temperature Recuperator;" DOE contract No. DE-FC01-80CS40330 (GTE Product Corp.) 1983.

3. J. W. Berklie and R. A. Penty, "High Temperature Recuperator Test;" DOE Contract No. DE-ACOA-79CS-40257 (Hague International) 1980.

4. M. G. Coombs, D. M. Kotchick, H. J. Strumpf, "High Temperature Ceramic Recuperator and Combustion Air Burner Programs," GRI-82/0015 (Garrett-AiResearch) 1982.

5. R. E. Womack and D. K. Stafford, "High Temperature Burner Duct Recuperator Program;" unpublished draft annual report (Babcock & Wilcox) 1982.

6. F. Rudloff, "High Temperature Counter Flow Recuperator;" DOE/CS/12077 DOE contract No. DE-AC07-80ID/2077 Final Report (Uintex Corp.) 1981.

7. M. E. Ward et al., "Development of a Ceramic Tube Heat Exchanger with Relaxing Joint;" Final Report DOE Contract No. EF-77-C-01-2556, FE-2556-30 (Solar Turbines International) 1980.

8. M. E. Ward and A. H. Campbell, "Ceramic Heat Exchanger Technology Development;" Final Report DE-4898-F, DOE Contract DE-ACOI-80ETB712. (Solar Turbines International) 1982.

9. A. Prasad and J. K. Jasti, "Advanced Regenerative Heat Recovery System;" Annual Report GRI-80/0115, GRI Contract No. 5080-342-0394, (Midland-Ross) 1982.

10. P. M. Wikoff et al., "High Temperature Waste Stream Selection and Characterization;" EGG-SE-6349, (preliminary report) August 1983.

11. M. Merrigan, "A Heat Pipe Exchanger Model for High Temperature Recuperators," CUBE Symposium, October 22-24, 1980, Lawrence Livermore National Laboratory, LA-UR-80-1985.

12. M. Merrigan, "Heat Pipes for High Temperature Waste Heat Recovery," 8th Energy Technology Conference, March 9-11, Washington, D.C., LA-UR-80-2431.

13. M. Merrigan and D. S. Sandstrom, "Ceramic Heat Exchangers, Manufacturing Techniques and Performance," WESTEC Symposium, Los Angeles, CA, LA-UR-81-1643, 1981.

14. M. Merrigan, W. Dunwoody, L. Lundberg, "Heat Pipe Development for High Temperature Application," 4th International Heat Pipe Conference, September 17-22, 1981, London, England, LA-UR-80-3198.

15. H. J. Strumpf and W. S. Miller, "Ceramic Heat Pipe Recuperator Study," Garrett-AiResearch Manufacturing Company of California, Torrance, CA, (under contract to LANL, Los Alamos, NM, 79-16480) May 12, 1980.

4. Evaluations

As stated earlier, the main purpose of this report is to evaluate and compare EG&G developed concepts against state-of-the-art and emerging heat recovery systems. This section describes how the evaluation was undertaken and details of the various evaluation components. The concepts were all evaluated against the design criteria developed and described in Section 2 of this report. These criteria include durability as it is affected by corrosion (Section 4.1.1) (some concepts are eliminated from further evaluation in this section because of the unsuitability of materials of construction to the waste streams of interest) and as it affects the amount of maintenance (Section 4.1.2), performance as exhibited by fouling (Section 4.3.1) and as it affects complexity of operation (Section 4.2.2), and as exhibited by energy savings (Section 4.2.3), and material and fabrication costs (Section 4.3). The concepts were also evaluated based on a typical economic analysis. The results were compared with the above design criteria evaluation. Results of the separate evaluations were then compiled and a consensus among the evaluators was obtained. The consensus results are summarized in Section 4.4.

In this evaluation, it was assumed that the two EG&G concepts would work within the design constraints even though there are still many unknowns, but because of these unknowns, these concepts are somewhat downgraded. Since the EG&G concepts are barely developed and information on emerging and state-of-the-art designs was obtained from the developer (although they were thoroughly reviewed for inconsistencies), the following evaluation results can only be accepted on a gross comparative basis.

4.1 Waste Streams Corrosive Effect on Material Durability

The ceramic materials which have been used in the emerging and state-of-the-art concepts are silicon carbide, cordierite, and refractory brick compositions. The EG&G concepts will use fibers of zirconia alumina, or silicon carbide. These materials, where exposed to high-temperature corrosive environments, may fail due to degradation of their structural integrity over time. This reduces the service-life of the recuperator. The characteristics of the waste heat stream (which were discussed in Section 2 and which have not been extensively reported[1,2,3,4]) determine to the greatest extent this service-life. In detail, the factors or variables which effect the reliability of a ceramic recuperator in these waste heat streams include:

- Temperature

- Degree of thermal cycling

- Presence of particulates

- Amount and composition of fluxes (such as in an aluminum remelt furnace)

- Amount and composition of batch carry-over (such as in a glass furnace)

- Amount and type of contaminants in the fuel

- Amount and composition of other contaminants such as topping compounds (such as in a steel soaking pit)

- Mechanical stresses inherent to the recuperator design. The durability of recuperator materials in terms of these variables will be discussed for the waste streams of interest.

4.1.1 Heat Exchanger Ceramic Corrosion Evaluation

Ceramic materials are generally considered corrosion resistant, although little effort to document the effects of corrosive atmospheres on the structural reliability of these materials has been put forth. Available corrosion data have been reviewed and form a basis for the following discussion. The actual corrosion testing completed to date generally involves coupons being placed in selected waste streams, without mechanical stress, and then subsequently strength tested at room temperature after exposure. Following is a discussion of the corrosion/durability aspects of each of the candidate ceramic materials.

4.1.1.1 Silicon Carbide (SiC). Silicon carbide is the primary candidate material for high-temperature recuperation. Several types of SiC are commercially available and their performance in varied environments is discussed herein.

1. Reaction-Sintered or "Siliconized" SiC. The materials in this category are produced by mixing silicon metal with carbon along with SiC grains, and then firing. The firing converts much of the free silicon and carbon to form a secondary phase of finely dispersed SiC grains. The microstructure is composed of a large primary SiC phase, a small grained (reaction-sintered) SiC phase, free silicon, and porosity. Carborundum and Norton Companies both produce a reaction-sintered SiC material, called KT and NC-430, respectively. The tube-in-tube concept and the ball and socket tube-in-shell concept both use this material.

   The primary corrosion problem associated with reaction-sintered SiC is alkali attack of the free silicon phase. It has been shown that glass atmospheres, which contain high amounts of alkali, react with this material to form a glassy slag on the surface. This glassy slag is a low viscosity glass that is produced by the combination of alkali and the free silicon phase. Also, it has been evidenced that cristobalite, a high temperature form of SiO_2, is also present in the surface slag. Furthermore, the depletion of free silicon from the matrix causes porosity to develop in the material. This porosity reduces the ultimate strength of siliconized-SiC in environments containing large amounts of alkali. Also, fairly rapid reactions occur at lower temperatures (2237°F) and much slower corrosion occurs at higher temperatures (>2570°F) with the degradation of the free silicon phase having little apparent effect on the bulk corrosion rate.[5]

2. **Recrystallized or Self-Bonded SiC.** This material is produced by sintering fine-grained silicon carbide without sintering aids or secondary phases. SiC crystals will tend to grow and stick together without shrinkage. The result is a SiC skeleton which is rather porous and therefore permeable. Norton NC-400 is a good representative of recrystallized SiC. This material was primarily intended for coating with an impervious skin. Although it is not exactly like the material used in the spring loaded tube-in-shell recuperator, it can be classified as such because of its extensive porosity.

   Recrystallized SiC has been shown to dissolve and possibly reprecipitate at the interface between the silicon carbide and the amorphous coating in a glass furnace environment.[5] The amorphous coating will penetrate easily into the open porosity present in this material. A reduction in strength is evident due to the bulk corrosion that occurs.

3. **Sintered α-SiC.** Sintered α-SiC is produced by mixing sintering aids such as 0.5% boron and 0.3% carbon with fine grained (<1μm) silicon carbide and then sintering. This material is very dense and exhibits the best mechanical properties of all silicon carbides currently available. Carborundum presently produces this material and General Electric is working on a similar form. The compliant seal tube-in-shell recuperator employs this material in an area of the recuperator which will be exposed to the greatest corrosion.

   Sintered α-SiC is the most corrosion resistant SiC material produced to date. However, since it is a nonoxide material, it relies on a silicon dioxide protective layer which develops at high temperature. Below 2192°F, this protective layer is a glass and above 2192°F, the glass will crystallize to cristobalite, which is still protective. The rate of oxidation of this nonoxide material is controlled by the silica film. Therefore, anything which alters this film will indeed affect the oxidation or corrosion rates. Alkali attack will create a low melting glassy slag on the surface of the SiC which will degrade this material by acceleration of the oxidation rate.

4. <u>Nitride-Bonded SiC</u>. This material is produced by mixing SiC grains with free silicon and then reacting the compact powder with nitrogen to give Si_3N_4 as the bonding medium. Carborundum, NGK, and Norton are all experimenting with this material. Carborundum calls their material REFRAX 20, NGK C/75C and Norton NC 136.

Nitride-bonded SiC (NGK) material has been exposed to aluminum remelt and steel soaking pit environments by Garrett-AiResearch for a relatively short period of time. There was no evidence of surface recession due to corrosion, although this material was extensively cracked throughout the total wall thickness with many cracks perpendicular to the surface. The low strength of this material after exposure indicates poor durability qualities.

4.1.1.2 <u>Cordierite</u>. Cordierite has an ideal composition of 2 MgO 2 $Al_2O_3 \cdot 5SiO_2$ and is a good candidate for high-temperature recuperation because of its excellent thermal shock resistance due to a very low thermal expansion coefficient. G.T.E. has successfully used this material for relatively high-temperature recuperation in clean waste streams. It has been used in waste heat streams as high as 2500°F (1371°C). High sodium and calcium contents present in fluxing compounds will form low melting aluminosilicates which will yield poor structural stability of cordierite at temperature. The possibility of coating this low expansion material with a more chemically resistant material could improve its chemical compatibility with dirty waste streams.

4.1.1.3 <u>Refractory Brick Compositions</u>. Several refractory brick compositions have been considered for heat recuperation in a glass furnace environment. The majority of corrosion work on these materials has been performed by the Unitex Corporation in Salt Lake City, Utah. They found three materials to withstand a soda-lime glass environment over the entire temperature range. These materials and manufacturers are:

- Corhart--Unichrome

- Corhart--X-81

- Glass and Ceramics International--ECP-3.

Details of these materials can be found in Reference 6. In general, all these refractories are alumina or magnesia based with large amounts of chromium oxide. It appears that the chromia helps the refractory material withstand the high temperature slag environment since only chromia-bearing materials withstood the test conditions. However, this report states that fabrication difficulties have made certain designs unacceptable.

4.1.1.4 **Alumina (Al_2O_3)**. Alumina is a very corrosion resistant and wear resistant material; such resistance is dependent on the purity content. Although it has a lower thermal conductivity than SiC, its chemical stability in a corrosive waste stream is better. The primary problem with Al_2O_3 is poor thermal shock resistance. Therefore, thermal cycling of furnace with an alumina recuperator is not suggested. The presence of small amounts of certain impurities in an alumina ceramic, will affect the high temperature mechanical and chemical properties which are desirable. Especially SiO_2 will cause an alumina ceramic to creep at elevated temperatures. High purity Al_2O_3 ceramics, where thermal cycling is not prevalent, could be used in very dirty waste streams at high temperatures.

4.1.1.5 **Zirconium Oxide (ZrO_2)**. Zirconium is a very good heat resisting material that has been considered for high temperature heat recovery systems.[5] The material was rejected during preliminary evaluations due to cracking resulting from exposure testing. This cracking is caused by a phase transformation, tetragonal to monoclinic, which occurs at 1170°C and is accompanied by a volume increase on transformation. This volume change during thermal cycling has made pure zirconia unsuitable for high temperature recuperation systems. However, the addition of calcia or yittria or magnesia has been shown to eliminate this transformation.[7] This zirconia-based material is now called stabilized zirconia and is a valuable refractory. Stabilized zirconia has not been evaluated by corrosion testing as a potential candidate for high temperature heat recuperation. Therefore, it is now being considered in this report for application in the conceptual designs. Further testing is required in order to make a good evaluation of this material.

4.1.2 Summary of Corrosion Evaluation of Ceramic Materials

In summary, based on the available corrosion information of materials exposed to different waste streams, the following list has been assembled to show the best corrosion resistant materials in each waste stream:

1. Steel Soaking Pit--Reaction sintered SiC, sintered α-SiC, and stabilized zirconia

2. Aluminum Remelt--Sintered α-SiC and stabilized zirconia

3. Glass Melting--Alumina and stabilized zirconia.

Steel soaking pit environments are the least corrosive of all three waste streams and reaction sintered SiC and sintered α-SiC are the best candidates for this application for corrosion resistance and thermal cycling. Stabilized zirconia might also be appropriate. Aluminum remelt environments present lower temperatures but increased alkali contents due to fluxing; therefore, sintered α-SiC or stabilized zirconia are the best candidates although stabilized zirconia is an unproven material. Glass manufacturing environments present the most difficult problem because of their extreme high temperatures and very corrosive atmospheres. Therefore, alumina is a good candidate if the glass furnaces are continually operating. Otherwise, stabilized zirconia should be tested for this potential application.

Although fiber structures have a great potential for successful use in high-temperature recuperation, it is very difficult to compare corrosion of heat exchangers made from monolithic components to fiber structures at this stage of development. The corrosion resistance of a material is related to how well its protective system holds together during exposure to a waste stream. Since no corrosion information has been gathered on the corrosion resistance of fibers to various waste streams, a comparison of monolithic components to fiber structures is impossible.

The durability of a material as affected by corrosion has a direct effect on the replacement schedule of parts and the unit. This is discussed in more detail in Section 4.2.2.

REFERENCES

1. H. H. Blaw, "Trends in Continuous Glass Melting Technology," pp. 1-5 in Advances in Glass Technology, G. E. Rindone (ed.), The American Ceramic Society, Columbus, OH, 1962.

2. B. J. Kirkbride, "Chemical Changes Occurring During the Cooling of Hot Gases from Flat Glass Furnaces," Glass Technology, 20[5] (1979) 174-80.

3. D. M. Sanders, M. E. Wilke, S. Hurwitz, W. K. Haller, "Role of Water Vapor and Sulfur Compounds in Sodium Vaporization During Glass Melting," Journal American Ceramic Society, 64[7] (1981) 399-400.

4. P. M. Wikoff et al., "High Temperature Waste Stream Selection and Characterization," EGG-SE-6349, August 1983.

5. G. W. Weber and V. J. Tennery, "Materials Analyses of Ceramics for Glass Furnace Recuperators," ORNL/TM-6970, November 1979.

6. F. Rudloff, High Temperature Counter Flow Recuperator, Final Report, May 1981.

7. A. H. Heuer, "Alloy Design and Partially Stabilized Zirconia," Advances in Ceramics 3, Science and Technology of Zirconia, pp. 98-115, 1981.

4.2 A Comparison of the Performance of Recuperators

The performance of the state-of-the-art, emerging, and proposed recuperators as affected by fouling (especially in replacement and cleaning maintenance), fuel savings, and operating considerations are discussed in the following sections.

4.2.1 Recuperator Fouling Considerations

The three energy intensive process industries selected for recuperation in this study all result in extremely dirty and corrosive furnace exhaust environments. The glass melting waste stream is the most severe, followed by the aluminum remelting waste stream and the steel soaking streams. Many of the exhaust constituents are the same for all the exhaust streams; however, glass and alkali carryover increases the potential for fouling in the glass melting exhaust. (Fouling is defined in this report as the build-up of deposits on the heat exchange surfaces.) The exhaust characteristics of the three types of furnaces were summarized in Section 2.1.

The effect of fouling on heat exchanger design is twofold. In an exhaust recuperator, perhaps the most important is the effect of the foulant layer on the pressure drop of the unit since the available head is usually very low on the waste gas side. The build-up reduces the flow area and increases the effective surface roughness, two factors which increase the pressure drop. It is expected that thin slag layers will not present a major problem; however, excessive slag thicknesses and the lower density sulphate-type layers will be more detrimental. Periodic cleaning will alleviate the latter problem.

The second detrimental effect of the foulant layer is an increase in the overall thermal resistance between the two gas streams. This can severely reduce the performance of the unit. Here too, thin slag layers should not present a problem, but heavy slag deposits or significant low conductivity sulphate-type build-up will be a problem. If the deposit is cleanable, the solution will be to slightly overdesign the unit to accommodate a certain amount of fouling and then to determine a cleaning schedule to return the unit to its original performance level. The cleaning frequency (presumably air lancing) will need to be determined by test since it is totally application dependent.

4.2.1.1 Fouling Potential In Glass Melting Furnaces. In the current brick regenerators (checkers), most of the large particles and batch carryover (up to 20%[4] of the batch is carried out into the waste stream) are left in the relatively large passageways and are removed manually at large intervals. In the current, more compact units, it is expected that fouling and plugging will be even more of a problem because of the smaller passageways in these units. References 1 and 2 give a detailed description of the mechanisms of fouling and corrosion in the various regions of the glass furnace exhaust. These mechanisms are summarized below.

Problems caused by the exhaust products are twofold. First, the products can either cause or enhance the degradation of heat exchanger surfaces by corrosion, chemical reaction, etc. This chemical attack is discussed in Subsection 4.2 of this report. Second, deposition and condensation of certain exhaust constituents can occur which will degrade the heat exchanger performance both by lowering the overall heat transfer conductance of the unit and by increasing its flow resistance and hence, pressure drop. The latter will result in a loss of combustion air to the furnace, a reduction in fuel burning efficiency, and quite possibly an accelerated build-up from carryover products. The build-up of deposits on the heat exchanger surfaces is termed fouling. Although corrosion can accelerate fouling layer build-up, the coupling will not be considered in this evaluation.

Major foulants in glass furnace exhaust are particulate carryover from the furnace (Na_2O, SiO_2, Al_2O_3, V_2O_5, CaO, and cullet dust), eroded bricks, and sodium sulphate (Na_2SO_4) formed by combination of Na_2O and SO_x gases. The particulate matter in a conventional furnace tends to settle out in the brick checkerwork, requiring periodical cleaning. This fouling layer is described as a hard, white slag formed by a eutectic of Na_2SO_4, $MgSO_4$, and $CaSO_4$. It is expected that the white slag which contributes to the checkerworks plugging will occur at the appropriate temperatures observed in the recuperators. These glassy slags will be difficult to remove from relatively fragile recuperator surfaces.

The remaining Na_2SO_4 solidifies at approximately 1600°F into a fluffy white powder which will then be deposited on downstream surfaces. Reference 2 showed that the particulate matter exiting the checkers primarily consists of 200 ppm of Na_2SO_4 particles generally under 2 μ in diameter; the remainder was trapped in the checkers. However, in the present application, all the particulates will enter the recuperator. These particles will build up on the heat exchange surfaces but are known to be cleanable (air lance, steam lance, water wash) when deposited at the elevated temperatures in this application. Since these recuperators will not be operating below a waste gas temperature of 800°F (except during transient startup and shutdown operation), the hard scale formation and acid condensation problems characteristic of low-temperature application (waste gas less than 600°F) should not be significant.

4.2.1.2 *Fouling Potential In Steel Soaking Pits*. Steel soaking pit exhausts are in the 2450°F range once the units are brought up to operating temperature. Fuels used include natural gas, Number 2 fuel oil, and infrequently, Number 6 fuel oil. As in the glass melt furnaces, the foulants from the fuel combustion process are fuel dependent with natural gas being the cleanest. Number 2 fuel oil has up to 1% sulphur, thus increasing SO_x concentration in the exhaust. Number 6 fuel oil has 2 to 3% sulphur and vanadium present, resulting in significantly higher SO_x formation (hence Na_2SO_4).

An analysis of deposits from flue gases from steel soaking pits[1] shows little sulphate deposition; rather the predominant compounds are iron oxides carried from the furnace as molten particulates. SiO_2 and Al_2O_3 are also observed in the flue gas deposits. It is possible that this furnace was fueled by natural gas with a low sulphur content. It would then be expected that use of Number 2 and particularly Number 6 fuel oil would increase the SO_x and sulphate formation and the latter would produce V_2O_5, much like in the glass furnace environment.

Thus, the foulants anticipated from the steel soaking pit furnaces include iron and refractory oxides created at the higher temperatures in the furnace and probably dropping out early in the recuperator, and Na_2SO_4 and other sulphates depositing in a cleanable powder form below 1600°F. It may be difficult to remove the oxides since they will be much the same as the glassy slags in the glass plant exhaust.

4.2.1.3 <u>Fouling Potential In Aluminum Remelt Furnaces</u>. The aluminum remelt furnace exhaust considered in this study is at about 2100°F, somewhat cooler than that from the other furnaces. Air preheat levels will be somewhat lower because of this. Foulants from the fuel combustion process are dependent upon the type of fuel used. Sulphates and vanadium content of the exhaust increases as lower grade fuel oils are used. Number 6 fuel oil is more common in the aluminum remelt furnaces than in the steel furnaces.

The fluxes used in the aluminum remelt furnaces will result in highly corrosive halide vapors and, at lower temperatures, corrosive sulphate formation. Relatively cleanable sulphate particle deposition will occur at lower temperatures (below 1600°F for Na_2SO_4). Difficult-to-remove slagging deposits of oxides (predominantly Al_2O_3) will occur in the higher temperature regions of the recuperator. Although the amount of slag is not known, the buildup will be considerably less severe than in the glass melting furnace.

4.2.1.4 <u>Fouling Summary</u>. In summary, the foulants for each of the furnaces can be considered in two general categories. In the higher temperature regions of the heat exchangers, it will be difficult (if not impossible) to remove the slag layer of oxides and sulphates. This will be very severe in the glass plant exhaust because of the higher temperatures and greater batch carryover. In the aluminum remelt and steel soaking pit furnaces, oxides will deposit on the heat exchanger surfaces but the rate should be considerably slower than the build-up in the glass plant exhaust. Very little of the slag deposits will be removable.

At somewhat lower temperatures (1600°F for the major precipitant, Na_2SO_4) friable sulphates will solidify and deposit on the surfaces.

Condensing liquids can sometimes cause a hardened sulphate layer on the exchanger surface; otherwise, the deposit is predominately a fluffy powder and can be removed by soot blowers.

Acid condensation will not be a problem during normal operation since the stream is to be maintained above 800°F. Startup and shutdown cycles will result in small amounts of acid condensation; the effect of this needs to be considered.

4.2.1.5 <u>Prevention of Fouling Problems</u>. The amount of slag formation is dependent upon the type of process, fuel, furnace construction, temperature, and additives. However, there is little doubt that it will occur and be impractical to remove. The glass plant exhaust, for example, is found to plug checkerwork passages with cross-sections 4 in. x 6 in. in a few years. Perhaps a heat exchanger configuration that could somehow drain the slag into a sump region could operate without major catastrophic plugging for a satisfactory time period. Otherwise, the application of these units will have to be restricted to an exhaust stream with a tolerable (to be determined by test) level of contaminants, or be able to use "throw-away" modules (not necessarily the entire unit).

The softer particulate deposition occurring in the cooler regions of the recuperator are, in general, removable by air lancing. Steam or water wash would also be effective, but there remains a problem of what to do with the highly corrosive acids formed from the sulphates. In general, they are much easier to handle if they are removed when dry. One problem with air lancing that needs to be resolved is that of placement of the air nozzles in the very hot exhaust stream (where the major precipitation occurs) and then the complications of blowing the foulant through the heat exchanger passages without plugging them. The shell-and-tube designs are somewhat more open in this regard. Air pressures of 50 to 100 psi are required for effective removal, so the strength of the thin recuperator walls will need to be considered.

4.2.1.6 <u>Evaluation of Fouling in Recuperator Systems</u>. In comparing recuperator systems with regard to fouling, it is assumed that all of the systems perform the same task. That is, they all have the same heat duty, flow rates, and gas stream inlet and outlet temperatures so that process conditions have no significant influence on the comparisons. Thus, the comparison reduces to the effect of geometrical differences on foulant accumulation and/or removal.

The EG&G woven tube-in-shell heat exchanger has several advantages incorporated into its design to limit the fouling. These advantages include the relatively large passages that should not readily plug, flexibility of design which allows for smooth contours to minimize dead regions where fouling occurs, and the possibility of a slag collection basin in shell. Some disadvantages are that the structure may not handle slag loads or loads from air lance jets. Also, it may be difficult to air lance both upstream and downstream sides of tubes deep in the bundle where most deposits occur.

EG&G's paper, finned-plate heat exchanger concept was evaluated and it was determined that this concept had the following advantages: contouring of the inlet may minimize fouling on inlet face; there are no significant eddy regions, like back faces of tubes, where foulants can collect, slag can be collected in an interstage unit between modules, and small passages and thin walls give good thermal performance. Disadvantages are that small passages are prone to plugging and air lancing blows deposits through the small passages.

The SiC compliant-seal, tube-in-shell concept has the advantage that a shell-and-tube type exchanger with the waste stream on the shell side have relatively large passages so they should not readily plug. This design has the following disadvantages: the metal stage may be susceptible to corrosive attack by foulants, and it may be difficult to clean both upstream and downstream sides of tubes deep in the bundle where most deposits occur.

The SiC, tube-in-tube concept has the same advantage and disadvantage as the tube-in-shell concept: i.e., large passages which do not readily plug and difficulty in cleaning deep in the bundle.

Evaluations   79

A SiC, ball-and-socket-joint, tube-in-the-shell design has the following features which tend to prevent fouling problems: tube-in-shell exchangers with the waste stream on the shell side have relatively large passages so they should not readily plug and the axial waste gas flow should make cleaning easier than cross-flow. A disadvantage with this design is that the long tubes (approximately 15 ft) may increase difficulty of cleaning.

A SiC heat pipe design has large passages that should not readily foul and includes the possibility of a cleanout door in the side of the heat exchanger duct for tube access. A disadvantage of the heat pipe concept is that heat transfer augmentation (fins, etc.) would be desirable, but would worsen fouling and cleaning problems.

An alumina chromia helical counterflow design will tend to lessen fouling problems because the waste stream passage is relatively large with few dead areas to accumulate particulates. Secondary flows due to the helix may sweep out particulates. However, the helical passage may be somewhat more difficult to clean than a straight passage and build-up may occur in the narrow corners of the passage.

All of the above factors have an affect on how often maintenance needs to be performed.

4.2.2 Maintenance

Maintenance of the recuperator system and the furnace with a recuperator installed is an important and significant factor which must be evaluated. This is because of the severe operating environment encountered in the three recuperator applications which may cause failure of recuperator components and necessitate frequent cleaning of tubes. (Durability considerations were discussed in Section 4.1.1 and fouling considerations were discussed in Section 4.2.1.) Maintenance will be divided into preventive maintenance (including inspection) and annual maintenance.

4.2.2.1 **Preventive Maintenance**. Preventive maintenance is "the utilization of planned and coordinated inspections, adjustment, repairs, and replacements in maintaining an industrial plant."[5] The key to a successful preventive maintenance program is a well defined and properly performed inspection program. Certain factors are evaluated when determining the frequency of inspections. These are:

- Number of operating cycles

- Individual equipment conditions

- Importance to plant operations.

Beyond the inspection process, which is intended to identify potential, major problems, other aspects of a preventive maintenance program are the adjustment, repairs, and replacements of equipment. Equipment adjustments is a broad subject including valve adjustments, lubrication of fans, motors, etc. Repairs in the sense of preventive maintenance refers to those minor corrections or modifications that can be accomplished while the system is between batches, i.e., an extended shutdown is not required. This may include valve replacement, instrument calibration and adjustment, or control system adjustment. Replacements are such things as seal replacement, valve packing replacement, etc.

The advantages of a well-defined preventive maintenance program are obvious. It is the responsibility of the plant operator to formulate and perform all the necessary functions of a preventive maintenance program. From a review of the specific recuperator designs the only obvious comment to be made is that as the recuperator system's level of complexity increases so does the complexity of the preventive maintenance program.

4.2.2.2 **Annual Maintenance**. Annual maintenance is herein defined as the major component repairs and replacements necessary to provide efficient operation of a process system or support system throughout its operating lifetime. For most recuperator applications, replacement and/or cleaning of heat transfer ceramic tubes is the major consideration with regard to annual maintenance. The time required for performance of annual maintenance activities is dependent on the design of the system and should thus be considered during the design phase. If the recuperators were built in module form, the cost and complexity of the maintenance would be decreased. Of course, maintenance frequency is an important consideration. From Reference 6, it was determined that the annual costs for equipment maintenance and repairs can be as high as 20% in cases where there are severe operating demands.

4.2.2.3 <u>Maintenance Evaluation</u>. Only tube-in-shell designs (such as Hague's CERHX SiC spring-loaded heat exchanger) have been commercialized; therefore, specific information on maintenance requirements only exists for these units. All of the other concepts are in various stages of conceptual development. Generally, maintenance requirements or programs are not established until a specific application is selected and the design is finalized. This makes it extremely difficult to evaluate the different concepts on the maintainability aspects. Certain aspects related to maintenance can be incorporated early in a design. These may include such things as access considerations for maintenance, addition of cleaning ports, easily removable modules/tube bundles, and the pitch of the tubes.

Table 4.1 is a summary of the annual maintenance assumed for each of the systems/concepts. It was assumed that there were no major differences in the preventive maintenance needs between any of the recuperator systems and that all preventive maintenance functions would be performed without the need for additional personnel or equipment. Table 4.1 also lists those aspects of the various recuperator designs identified as advantages or disadvantages with respect to maintenance. It is noted on several of the concepts that leakage is a potential problem. This is a concern with regard to maintenance, i.e., minimizing leakage in a severe environment.

A review of the different concepts and their required maintenance levels indicates that, generally, the aluminum remelt and glass remelt applications require more maintenance than the concepts applied to steel soaking pits. This is a result of the more corrosive waste streams in the aluminum and glass remelt plants.

TABLE 4.1 RECUPERATOR MAINTENANCE REQUIREMENTS

Design Configuration	Maintenance Application	Level	Advantages	Disadvantages
SiC Spring-loaded Tube in Shell	Aluminum remelt	Low (1-2 wk/yr)	Experience from several operating units. Small component size and design of ceramic tube loading method (spring-loaded) allows easy removal and replacement.	--
SiC Compliant Seal Tube in Shell	Steel soaking pit	Medium (4-5 wk/yr)	--	Large size (4 ft tall and 600 lb each) of modules. No cleaning ports identified. Requires complete replacement of module.
SiC Tube-in-Tube	Steel soaking pit	Medium (3-4 wk/yr)	Removable covers on recuperator modules. Individual ceramic tubes supported by a hanging bayonet mount.	Large module size. Tubes are on the order of 8 ft long.
Alumina-Chromia Helical Counterflow	Aluminum remelt Glass remelt	High (6-8 wk/yr)	Small, stackable modules. Removal capabilities designed into the concept.	Based on Terra-Tek's estimate of 8% module replacement (20 modules per year). This is a significant task.
SiC Ball in Socket Joint Tube in Shell	Steel soaking pit	High (5-6 wk/yr)	Individually mounted tubes with the ability to assemble (and disassemble) the module and replace the ceramic tubes or header.	Large unit size.

TABLE 4.1. (Continued)

Design Configuration	Maintenance Application	Level	Advantages	Disadvantages
Paper finned plate with ZrO	Steel soaking pit Aluminum remelt	High (5-6 wk/yr) High (8-10 wk/yr)	Can be designed as modular groups simplify the replacement.	There is an apparent difficulty in cleaning this type of unit because the thickness of the "paper" is minimal and would not withstand any abrasive cleaning. This is especially important when used in the more corrosive waste streams.
with SiC	Steel soaking pit	High (5-6 wk/yr)		
with Al O	Glass remelt	High (10-12 wk/yr)		
Woven Tube in Shell with ZrO	Steel soaking pit Aluminum remelt	Medium (4-5 wk/yr) High (6-8 wk/yr)	See (a) below	See (a) below
with SiC fiber	Steel soaking pit	Medium (4-5 wk/yr)		
with Al O	Glass remelt	High (8-10 wk/yr)		
Ceramic Heat Pipe	Steel soaking pit	High (8-10 wk/yr)	Capability to change individual heat pipes	--

a. None of the EG&G concepts have been developed to an extent whereas it is possible to evaluate the maintenance levels required. Design features to make maintenance simpler would be included in the design as it develops.

4.2.3 A Comparison of the Thermal Performance of Recuperator Furnace Systems

This section compares the thermal performance of representative emerging state-of-the-art and proposed EG&G recuperator concepts whose material of construction is most likely to reasonably survive in the three candidate waste heat streams as discussed in Section 4.1. The assumed thermal process conditions are given in Subsection 4.2.3.1. The emerging and state-of-the-art recuperators were originally designed for different uses so the performance was extrapolated to a common basis for each environment for purposes of comparison. The extrapolation method and resulting process conditions are also discussed in Subsection 4.2.3.1.

The recuperator concepts are compared on the basis of fuel savings (Btu/hr), local and global fuel savings (% of unrecuperated furnace), and second law efficiency (Section 4.2.3.2). The performance, based on an availability efficiency (effectiveness) for each case, is given to show the relative performance from the perspective of the second law of thermodynamics. Results of the performance calculations for the different recuperators are discussed in Subsection 4.2.3.3.

4.2.3.1 Recuperator Process Conditions for Each Application. Since information is scarce on determining the operating conditions for a generic furnace, operating conditions for each of the selected waste heat stream applications were obtained from the furnaces to which the emerging and state-of-the-art concepts have been applied and are summarized in Table 4.2. The inlet air temperature to the burners (the air temperature leaving the recuperator) was held to around 2000°F for the 2450°F steel soaking pit and the 2800°F glass melting furnaces because of limitations on the operating temperature of the burners. In the 2100°F aluminum remelting furnace, it was held to 1800°F so as to maintain a reasonable approach temperature difference. The load temperatures were used in the availability analysis and the operating cycle numbers are used in the economic analysis.

Table 4.3 gives recuperator process data analyzed for the steel soaking pit application. The soaking pit conditions from Table 4.2 are

TABLE 4.2. FURNACE OPERATING CONDITIONS

Application	Furnace Size (unrecuperated burner size) (Btu/hr × 10)	Air Inlet (°F)	Temperatures			Operating Cycle (% of time at high fire)	Turn Down Ratio
			Exhaust Gas Leaving Furnace (°F)	Load (°F)			
Steel Soaking Pit	22	100	2450	Varies linearly from 1800 to 2350		36	4:1
Aluminum Melting Furnace	20	100	2100	1400		80	5:1
Glass Melting Furnace	215	100	2800	Varies linearly from 1300 to 2700		100	not applicable

TABLE 4.3 PROCESS DATA FOR STEEL SOAKING PIT RECUPERATORS

	SiC Compliant Seal Tube in Shell Concept		SiC Tube in Tube Concept				Paper Finned Plate Concept (ZrO₂ or SiC) Predicted	
			Actual		Extrapolated (See Text)			
	Ceramic Unit	Metal Unit	Ceramic Unit	Metal Unit	Ceramic Unit	Metal Unit	Ceramic Unit	Metal Unit
Air								
Flow rate (lb/hr)	8,654	8,654	11,250	11,250	11,250	11,250	8,654	8,654
Temperature in (°F)	1,300	100	500	100	476	100	1,300	100
Temperature out (°F)	2,000	1,300	1,860	500	2,010	476	2,000	1,300
Pressure loss (in H_2O)	10.0	4.6	40.0	10.0	40.0	10.0	9.0	4.6
Exhaust Gas								
Flow rate (lb/hr)	9,114*	12,240*	11,814	11,814	11,814	11,814	9,114	9,114
Temperature in (°F)	2,450	1,500*	2,250	1,015	2,450	1,219	2,450	1,887
Temperature out (°F)	1,887*	758	1,015	695	1,219	846	1,887	972
Pressure loss (in H_2O)	3.2	1.3	1.0	0.5	1.0	0.5	2.5	1.3
Heat exchanger effectiveness	0.61	0.86	0.78	0.44	0.78	0.44	0.61	0.86

	Woven Tube in Shell Concept (ZrO₂ or SiC) Predicted		SiC Heat Pipe Concept			SiC Ball and Socket Joint Tube in Shell Concept		
			Actual	Extrapolated (See Text)		Extrapolated (See Text)		
	Ceramic Unit	Metal Unit		Ceramic Unit	Metal Unit	Ceramic Unit	Metal Unit	Actual
Air								
Flow Rate (lb/hr)	8,654	8,654	19,605	19,605	19,605	3,600	3,600	3,600
Temperature in (°F)	1,300	100	1,304	100	100	1,300	100	1,300
Temperature out (°F)	2,000	1,300	2,152	1,304	1,600	2,105	1,900	152
Pressure loss (in H_2O)	9.9	4.6	6.1	4.6	6.1	neg.	0.67	4.6
Exhaust Gas								
Flow Rate (lb/hr)	9,114	9,114	20,714	23,676*	23,676*	3,780*	3,600	3,600
Temperature in (°F)	2,450	1,887	2,125	2,450	1,500*	2,450	2,200	1,300
Temperature out (°F)	1,887	972	—	1,772*	719	1,809*	1,600	758
Pressure loss (in H_2O)	4.0	1.3	8.7	8.7	1.3	neg.	neg.	1.3
Heat exchanger effectiveness	0.61	0.86	0.74	0.74	0.86	0.70	0.67	0.86

*Cold air mixed with exhaust from ceramic unit to reduce inlet temperature to metal unit to 1500°F.

essentially those used in the design of the SiC compliant seal tube-in-shell recuperator. These values were also used for all of the EG&G designs (see Appendix F). All recuperators are made up of a ceramic unit for the high-temperature end and a metal unit, in series, at the low-temperature end except for the glass melting application. This was done so that all the recuperators could be compared on a more equitable basis with minimal changes in their designs, especially for the economic analysis. (An all ceramic unit is prohibitively expensive compared to ceramic metal systems which do the same job--see Appendix F for cost numbers.) Note that the exhaust gas conditions leaving the SiC compliant seal tube-in-shell ceramic recuperator are different from those entering this metal unit. This is the result of adding cold air to decrease the high temperature in the metal unit to 1500°F (the maximum operating temperature for the material).

The silicon carbide tube-in-tube concept was designed to operate with 2250°F exhaust gas instead of the 2450°F temperature used in this analysis. Because it would be nearly impossible to reoptimize these designs, it was decided to hold the heat exchange effectiveness constant when extrapolating the performance. This is approximately equivalent to using the same size unit because the heat capacity rates of both streams remain constant. The extrapolated results are shown in Table 4.3.

The heat pipe and the silicon carbide ball-in-socket joint tube-in-shell recuperator presented a difficult problem to put on a consistent basis. Each was designed as a single ceramic unit (the tube-in-shell unit for use on an industrial gas turbine). Their individual performance was not high enough to preheat the air to the designated temperature. More passes in a cross-flow arrangement would have been needed to increase the heat exchange effectiveness. If a metal unit, identical in performance to the SiC compliant seal tube-in-shell recuperator was used in conjunction with each ceramic unit, the desired condition could be reached. The extrapolated configurations in Table 4.3 show the process conditions with the two units in series. The increased effectiveness of this shell-in-tube recuperator over the actual tested values results from changing the heat capacity rate ratio to reflect operation which is much closer to stoichiometric conditions in the furnace than in the gas turbine which used large amounts of excess air.

For the aluminum remelting application, Table 4.4 gives the process data. Both EG&G designs were designed to the actual conditions (see Appendix F). The alumina chromia helical ceramic recuperator needed a metal recuperator on the cold end similar to the heat pipe and silicon carbide ball-in-joint socket tube-in-shell units for the soaking pit application. The performance and size of the silicon carbide compliant seal shell and tube unit were changed to make it consistent with the other designs. This allowed for less ceramic recuperator than in the actual unit by 43% to decrease the air preheat temperature from 1900°F to 1800°F.

The glass melting process conditions are given in Table 4.4. The EG&G concepts were designed to the given thermal conditions. It was felt that the best way to make the alumina chromia helical unit similar was to add enough length to achieve the design effectiveness. This was possible because of the countercurrent flow design of the helical concept. A 29% increase in the length and area of this concept was required.

Again, it is noted that only the EG&G concepts, for all three applications, and the SiC compliant seal tube-in-shell recuperator for the soaking pit were designed for the actual conditions identified in this report. Better performance, lower costs, and better material situations would probably have been obtained for all of the other units if they had been designed for the actual application.

4.2.3.2 **Fuel Saving and Performance Data**. Each of the recuperators listed in the previous subsection was analyzed at high-fire conditions to determine the amount of fuel it saved along with the added electrical power to account for the increased pressure loss in the system created by the recuperator. The recuperators were assumed to operate in a clean condition, and maintenance costs were used to account for fouling in the economics analysis. From an economic analysis standpoint, one is interested in comparing the fuel savings resulting from using a recuperator on a given furnace with an unrecuperated furnace with both furnaces

Evaluations 89

TABLE 4.4 PROCESS DATA FOR ALUMINUM AND GLASS MELTING

Aluminum Melting Application

	Woven Tube in Shell Concept (ZrO₂) Predicted		Paper Finned Plate Concept (ZrO₂) Predicted		SiC Compliant Seal Tube in Shell Concept				Alumina Chromia Helical Concept			
					Actual		Extrapolated		Actual		Extrapolated (See Text)	
	Ceramic Unit	Metal Unit	Ceramic Unit	Metal Unit	Ceramic Unit	Metal Unit	Ceramic Unit	Metal Unit	Ceramic Unit	Metal Unit	Ceramic Unit	Metal Unit
Air												
Flow rate (lb/hr)	9,990	9,990	9,990	9,990	9,990	9,990	9,990	9,990	16,200	16,200	16,200	16,200
Temperature in (°F)	951	100	951	100	751	100	951	100	1,067	100	100	100
Temperature out (°F)	1,800	951	1,800	951	1,911	751	1,800	951	1,707	1,067	2,000	1,600
Pressure loss (in H₂O)	22	3.8	15	3.8	19.6	3.8	19.6	3.8	12.5	5.2	16.1	12.5
Exhaust Gas												
Flow rate (lb/hr)	10,518	10,518	10,518	10,518	10,518	10,518	10,518	10,518	17,100	18,126*	17,100*	17,100
Temperature in (°F)	2,100	1,409	2,100	1,409	2,100	1,102	2,100	1,409	2,500	1,500*	2,100	2,500
Temperature out (°F)	1,409	752	1,409	752	1,102	—	1,409	752	1,238	815	1,584*	1,365
Pressure loss (in H₂O)	4.6	1.1	7.3	1.1	2.7	1.1	2.7	1.1	27.6	0.5	27.6	—
Heat exchanger effectiveness	0.74	0.65	0.74	0.65	0.86	0.65	0.74	0.65	0.62	0.65	0.62	0.70

Glass Melting Application

	Woven Tube in Shell (Al₂O₃) Predicted	Paper Finned Plate (Al₂O₃) Predicted	Alumina Chromia Helical Concept	
			Actual	Extrapolated (See Text)
Air				
Flow rate (lb/hr)	54,000	54,000	16,200	16,200
Temperature in (°F)	100	100	100	100
Temperature out (°F)	7,000	7,000	1,600	2,000
Pressure loss (in H₂O)	27.4	27.4	12.5	16.1
Exhaust Gas				
Flow rate (lb/hr)	57,000	57,000	17,100	17,100
Temperature in (°F)	2,800	2,800	2,500	2,500
Temperature out (°F)	1,365	1,365	1,238	1,365
Pressure loss (in H₂O)	7.3	27.6	35.4	—
Heat exchanger effectiveness	0.70	0.70	0.62	0.70

*Cold air mixed with exhaust from ceramic unit to reduce inlet temperature to metal unit to 1500°F.

producing the same output and producing the same output (in terms of heat transfer to the load) as the unrecuperated furnace. Therefore, the fuel saving calculations consisted of the following three items:

1. Calculating the heat transfer to the load (plus losses other than stack losses) in an unrecuperated furnace.

2. Calculating the fuel necessary to run the recuperated furnace with the same: (a) air/fuel ratio and (b) temperature of exhaust gas leaving the furnace (entering recuperator).

3. Comparing the answers to 1 and 2 to determine the saving.

Another method assumes that the energy input to the furnace, instead of the load, is constant. This assumption leads to a different load in the furnace but when expressed as per unit load, the answers are approximately the same. Because of the economic analysis format, it was decided that the method which assumes all concepts performed the same job was appropriate. Recuperators originally designed for different jobs will have different flow rates through the furnace to do the same job with a constant air/fuel ratio for this comparison.

Table 4.5 shows the air flow rate for all of the cases considered along with the fuel savings. The incremental power associated with adding the recuperator is also calculated. For the various systems, different values are given in the literature. For example, one recuperator is said to require no additional power because the installation into which it is being retrofitted had large enough fans to handle it already. To put all of the recuperation on the same basis, the following assumption was made to calculate the incremental power; a fan was assumed to be on the air stream before entering the recuperator which produced a pressure rise equal to the air-side pressure loss. Another fan was assumed to be in the exhaust gas stream after leaving the recuperator. This produced a pressure rise equal to the pressure loss in the exhaust gas side of the recuperator and the power requirement of these two fans was calculated. The fans and fan motors were assumed to have isentropic efficiencies of 60% and 90%, respectively.

TABLE 4.5 PERFORMANCE DATA

	Air Mass Flow Rate (lb/hr)	Fuel Saving[a] (Btu/hr x 10^{-6})	Incremental Fan Power Increase With Recuperator (kW)
Steel Soaking Pit (22 x 10^6 Btu/hr unrecuperated)			
SiC Compliant Seal Tube in Shell Concept	8654	7.02	5.9
SiC Tube in Tube Concept	8428	7.16	9.4
Woven Tube in Shell (ZrO_2 or SiC)	8654	7.02	5.8
Paper Finned Plate (ZrO_2 or SiC)	8654	7.02	4.8
Heat Pipe Concept (with metal recuperator from first application above)	8285	7.24	6.5
SiC Ball in Socket Joint Tube in Shell Concept (with metal recuperator from first application above)	8346	7.21	27.6
Aluminum Melting Furnace (20 x 10^6 Btu/hr unrecuperated)			
Woven Tube in Shell (ZrO_2)	9305	7.90	15.0
Paper Finned Plate (ZrO_2)	9305	7.90	14.8
SiC Compliant Seal Shell in Tube Concept	9305	7.90	12.3
Alumia Chromia Helical Concept (with metal recuperator from first application above)	9603	7.61	43.6
Glass Melting Furnace (215 x 10^6 Btu/hr unrecuperated)			
Woven Tube in Shell (Al_2O_3)	54000	123	189
Paper Finned Plate (ZrO_2)			173
Alumina Chromia Helical Concept (increased in length 29%)	54000	123	490

a. Averaged for duty cycle while operating (i.e., percentage of time at high and low fire). Does not include unit operating factor.

The results of the performance calculations are shown in Table 4.5 and are corrected for the actual duty cycle of the furnace assuming that the percentage of fuel savings was the same at low fire, as at high-fire conditions. Actually, it is slightly higher at low fire, but the magnitude of the energy saved is small at low fire. The fan power was assumed to vary as the square of the air flow rate. The differences in flow rates for the recuperators, due to the original design and required changes to make the units do the same job, are evident in the different incremental fan power increases required. The additional fan power required for the SiC ball-in-socket joint tube-in-shell concept in the soaking pit application and the alumina chromia helical concept in the aluminum and glass applications are particularly noteworthy.

Table 4.6 expresses the energy savings for the high-fire condition in a somewhat different manner. The first column expresses the fuel savings as a percentage of the unrecuperated furnace fuel flow rate. For this column, the fan power required was not included in the assessments. Another way to present the energy savings is to penalize each design additionally for the fuel used at a power plant to produce the electricity required to power its fan. Assuming a thermal efficiency of 30% for the power plants, resulting energy savings are shown in the second column, and described as global fuel savings. Note that for the present study this affects the results by only 1 or 2%. For each application, the percent savings for each type of recuperator is similar, mainly because the required performance in terms of air preheat temperatures is about the same.

The previous paragraphs present what might be called a "first-law-of-thermodynamics approach". The approach simply took the energy and made no distinction between high-grade (high-temperature) energy and low-grade (low-temperature) energy. An expansion of the performance evaluation of the maximum potential availabilities, as limited by the second law of thermodynamics, were calculated as described by Gaggioli and Petit.[7] The availability is defined as the maximum work which could be produced by an effort considering its interaction with the environment. An efficiency is defined by adding all the increases in availability and

TABLE 4.6 FUEL SAVINGS FOR VARIOUS RECUPERATORS AT HIGH FIRE CONDITIONS

	Fuel Saving (% of Unrecuperated Furnace)	Global Fuel Saving (% of Unrecuperated Furnace)
Steel Soaking Pit Application (22 x 10^6 Btu/hr unrecuperated)		
SiC Compliant Seal Tube in Shell Concept	58	57
SiC Tube in Tube Concept	59	58
Woven Tube in Shell Concept (ZrO_2 or SiC)	58	57
Paper Finned Plate Concept (ZrO_2 or SiC)	58	57
SiC Heat Pipe Concept	60	59
SiC Ball and Socket Joint Tube in Shell Concept	59	58
Aluminum Melting Furance Application (20 x 10^6 Btu/hr unrecuperated)		
Woven Tube in Shell Concept (ZrO_2)	47	46
Paper Finned Plate Concept (ZrO_2)	47	46
SiC Compliant Seal Tube in Shell Concept	47	46
Alumina Chromia Helical Concept	45	43
Glass Melting Furnace Application (215 x 10^6 Btu/hr unrecuperated)		
Woven Tube in Shell Concept (Al_2O_3)	67	66
Paper Finned Plate Concept (Al_2O_3)	67	66
Alumina Chromia Helical Concept	67	64

dividing the sum by the sum of all of the decreases in availability. For example, in a heat exchanger the cold fluid will increase in availability while the hot fluid decreases in availability.

For the furnace-recuperator system, the air-fuel-exhaust gas decreases in availability. In other words, the surroundings (everything but the furnace and load) decrease in availability because the exhaust gas has less potential to produce work than the entering air and fuel. The load in the furnace increases in available energy. The energy necessary to drive the fan motor is another availability decrease. The efficiency is taken as the increase in availability to the load divided by the sum of the decrease in availability of the air/fuel and exhaust gas and the electrical input. The work input is an important effect. If it is not included in the soaking pit assessment, the silicon carbide ball-in-socket, tube-in-shell recuperator and the silicon carbide tube-in-tube unit each have an efficiency of 73%. With the work included, tube-in-shell efficiency drops to 66% while the tube-in-tube efficiency decreases to 71%.

Table 4.7 shows the "second-law of efficiency" of the furnace-recuperator systems under two assumptions. First, the available energy input to the system is taken to be the change in availability of the air/fuel and exhaust gas stream as it flows through the system, plus the incremental work required. The work is referenced to availability change at the power plant to produce the work assuming an efficiency of the power plant. This assumes that the available energy stored in the exhaust gas will be utilized further in a bottoming cycle or cascaded to another process. Secondly, it was assumed that all of the available energy in the air/fuel mixture is expanded in the process and none is recovered. This in general does not change the relative ranking of the systems and may be a more realistic appraisal of what is done in practice.

4.2.3.3 **Discussion of Results**. The primary purpose of this analysis was to place the different systems on a common basis, compare their performance, and determine the operating parameters necessary for an economic analysis. This was done by using existing equipment or designs in conjunction with additional hardware (metal recuperator) deemed necessary for system performance. It should be reemphasized that most of the systems

TABLE 4.7 SECOND LAW PERFORMANCE OF RECUPERATOR-FURNACE SYSTEMS

	Second Law Efficiency	
	Assuming Energy In Exhaust From Recuperator Will Be Used	Assuming Energy In Exhaust From Recuperator Will Not Be Used
Steel Soaking Pit Application		
Base Case (Unrecuperated Furnace)	0.50	0.28
Furnace with Recuperator:		
SiC Compliant Seal Tube in Shell Concept	0.70	0.64
SiC Tube in Tube Concept	0.71	0.65
Woven Tube in Shell Concept (ZrO_2 or SiC)	0.70	0.64
Paper Finned Plate Concept (ZrO_2 or SiC)	0.70	0.64
SiC Heat Pipe Concept (with metal recuperator from first application above)	0.71	0.66
SiC Ball and Socket Joint Tube in Shell (with metal recuperator from first application above)	0.66	0.62
Aluminum Melting Furance Application		
Base Case (Unrecuperated Furnace)	0.51	0.33
Furnace with Recuperator:		

TABLE 4.7. (continued)

	Second Law Efficiency	
	Assuming Energy In Exhaust From Recuperator Will Be Used	Assuming Energy In Exhaust From Recuperator Will Not Be Used
Woven Tube in Shell Concept (ZrO_2)	0.64	0.55
Paper Finned Plate Concept (ZrO_2)	0.64	0.55
SiC Compliant Seal Tube in Shell Concept	0.64	0.56
SiC Helical Concept (with metal recuperator from first application above)	0.61	0.57
Glass Melting Furnace Application (215 x 10^9 Btu/hr unrecuperated)		
Base Case (Unrecuperated Furnace)	0.40	0.18
Furnace with Recuperator:		
Woven Tube in Shell Concept (Al_2O_3)	0.63	0.52
Paper Finned Plate Concept (Al_2O_3)	0.64	0.52
SiC Helical Concept (with length increased by 29%)	0.59	0.49

were not optimized for the actual application. The only systems designed for the actual applications were the silicon carbide compliant seal tube-in-shell concept for the steel soaking pit and the EG&G concepts for all three environments.

Because the systems were made to operate under similar conditions, there are no large differences in the "first- or second-law efficiencies". The "global first-law efficiencies" vary only 4% for a given furnace application. The "second-law efficiencies" for the steel soaking pit vary only 4% with the silicon carbide ball-in-socket joint tube-in-shell recuperator system being the lowest. This is a result of the high pressure drop on the air side of this unit since it was designed for a gas turbine recuperator and, in that application, could allow a 5-psi drop on the air side and thus, should not be penalized in this comparison since it would be designed for lower pressure drop in a furnace application.

For aluminum remelting furnace applications, the silicon carbide compliant seal tube-in-shell and the EG&G concepts unit appear to be slightly better from a second-law point of view than the alumina chromia helical concept. For glass melting furnaces, the EG&G concepts are slightly better than the alumina chromia helical design.

In general, for this comparison, the concepts show approximately the same fuel savings. (The fuel savings could be increased by preheating the air to a higher temperature.) The size and the cost of the unit to do this job are the parameters which will have significance in ranking the concepts. This is discussed further in Section 4.3. For example, the amount of pressure drop which increases the needed fan power as reflected in the global fuel savings, and the resulting second law effectiveness needs to be balanced against use of heat-transferred augmentation devices which decrease the size of the heat exchanger required, but which may increase capital costs. The effect of dilution air on performance would also impact the size and the cost of the unit. Cost considerations are discussed in more detail in Section 4.3.

4.2.4 Operating Considerations

Any high-temperature recuperator installed in either a steel soaking pit furnace, an aluminum remelt furnace, or a glass melting furnace will require a certain amount of operation support. Some recuperator operating factors which will be addressed in this section include required control systems, operating labor (e.g., plant operators, equipment operators, etc.), and utilities (i.e., electrical power). The operation of a recuperator system is divided into three areas of discussion: (a) the operating control systems required, (b) operating labor required, and (c) utilities required.

4.2.4.1 <u>Recuperator Control Systems</u>. The operations of recuperator systems and recuperated furnaces are reasonably straightforward and simple. Three automatic control systems that may be required for proper functioning of the recuperator and process furnace are a modification of the furnace pressure control system, an expanded temperature monitoring system, and a new blower (recuperator) pressure control system. In addition, safety systems and interlocks may be required.

Figures 4.1, 4.2, and 4.3 present concepts proposed by Garrett-AiResearch Manufacturing Company for a furnace pressure control system, a temperature monitoring system, and a blower pressure control system, respectively.[8]

It should be noted that only the blower pressure control system is dedicated totally to the recuperator system. A furnace pressure control system and a furnace temperature monitoring system of sorts would be required even in an unrecuperated furnace application. The complexity of these systems does not significantly increase with the addition of a recuperator.

4.2.4.2 <u>Operating Labor</u>. It is assumed that the operating manpower required for the successful and efficient operation of a recuperator is minimal. The inclusion of modern control and monitoring systems significantly reduces the efforts required from an operating staff. Labor activities required beyond those associated with the furnace may include

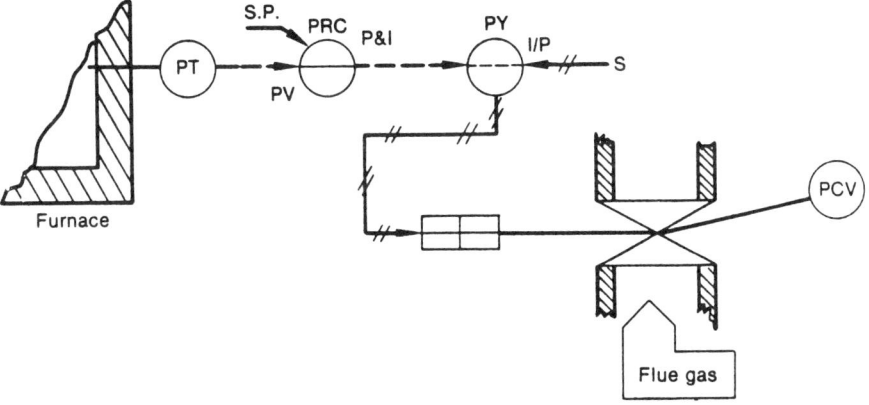

Figure 4.1 Schematic of furnace pressure control system.

100   Ceramic Heat Exchanger Concepts and Materials Technology

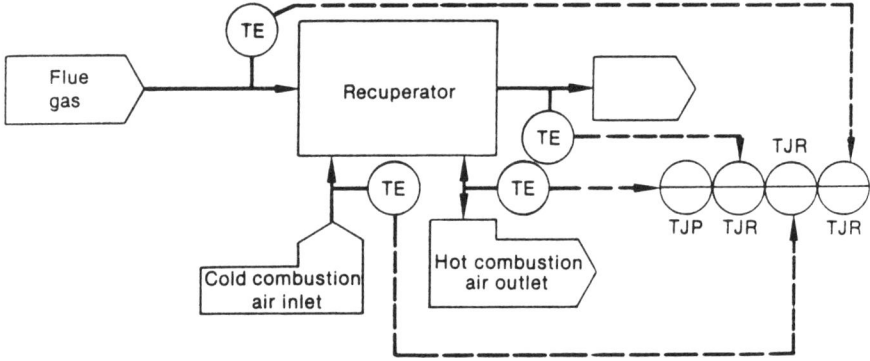

Figure 4.2   Schematic of temperature monitoring system.

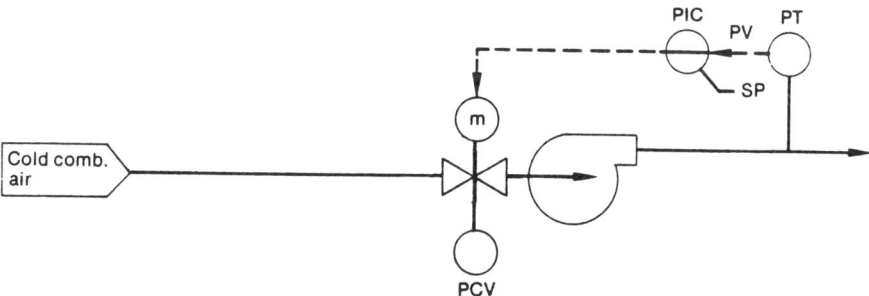

Figure 4.3 Schematic of blower pressure control system.

such things as maintaining status-log books on the recuperator system and performing minor equipment adjustments. The additional manpower required is assumed to be approximately 10 man-hours per week. For a furnace operating 24 hours per day, five days per week, this amounts to less than a 5% increase in labor costs above the costs incurred for an unrecuperated furnace system.

4.2.4.3 Operating Utilities. Electrical power must be supplied to an assortment of electrically driven recuperator components. These would include motors (for fans and valves) and instrument recorders, etc. The total cost of electricity for the recuperator system is dependent mainly on the fan size requirements and may range from several hundred dollars to several thousand dollars. Tables 4.5 and 4.9 show the anticipated electricity requirements for each application.

4.2.4.4 Operational Evaluation. Of the five commercially developed recuperators, the two EG&G recuperator concepts, and the ceramic heat pipe concept, only Hague's CERHX tube-in-shell heat exchangers have been commercialized and installed in furnaces. Within the design limitations of each commercial recuperator, their operating experience has been good. The recuperator design has been developed to the extent that early problems have been corrected by design and/or operational changes. All the other recuperator concepts are passive and do not require extensive operational support such as control systems, manpower, or electrical power.

Table 4.8 is a summary of the operational support required for each of the concepts. The operational support is delineated into three relative levels: low, medium, and high. It must be realized that even a high rating does not indicate that the amount of operational support required is excessive when compared with other process equipment systems.

4.3 Economic Analysis

The objective of the economic analysis is to compare the EG&G concepts with existing and other emerging recuperator technologies. The comparison was performed for typical application in each of three industries: a 22 MMBtu/hr steel soaking pit, a 20 MMBtu/hr aluminum remelt furnace and a

TABLE 4.8. RECUPERATOR OPERATIONAL SUPPORT

Design Configuration	Operational Support Needed
SiC Spring-loaded Tube-in-shell[a]	Low to medium
SiC Compliant Seal Tube-in-shell	Low
SiC Tube-in-tube	Low
Alumina Chromia Helical counterflow	Medium
SiC Ball in Socket Joint Tube-in-shell	Low
Paper Finned Plate	Low
Woven Tube-in-Shell	Low
Heat Pipe	Medium

a. Hague has recuperators in commercial operation. Therefore, the operational support required for this application is better and more extensively defined. This may account for the slightly higher support level identified herein.

215 MMBtu/hr glass melting furnace. For each application, the EG&G concepts were compared against other recuperator designs suitable for that application. The competing recuperators were ranked by economic standards used by private industry. Payback, rate of return, and net present value were calculated for each installation on the basis of after tax cash flows. The methodology employed is explained in Section 4.3.3. The costs used for these analyses are discussed in Section 4.3.1 and in more detail in Appendices A and B. The sensitivity of the economic results to the input parameters is discussed in Section 4.3.3.

At an early stage in development, it is difficult to estimate costs, energy savings, and the replacement period for new recuperator designs or for designs which were designed but not tested for the same situation, but which need to be adjusted for comparison on a common basis. Therefore, this economic analysis should be viewed on a relative basis only, so, although specific costs are developed in this section, their significance is in their relative magnitude, not absolute magnitude, in order to determine if the new proposed concepts offer an economic advantage relatively.

4.3.1 Cost and Energy Savings Estimates

The economic analysis is based on the cost and energy savings estimates shown in Table 4.9. The capital cost estimate was derived by sizing each recuperator to a base case application (i.e.: 22 MMBtu/hr steel soaking pit, 20 MMBtu/hr aluminum remelt furnace or 215 MMBtu/hr glass furnace). The details are included as Appendix G. The capital costs given include installation costs and control systems. The annual maintenance and periodic replacement costs in Table 4.9 are, for the most part, based on a percentage of the original capital cost. The replacement period is EG&G's estimate of each recuperator's life in each application based on corrosion and fouling of heat transfer surfaces. The annual operating cost estimate is the same for all of the recuperators. It is based on EG&G's estimate of the overall difference in annual operating costs between a facility with and without a recuperator installed. All of the cost estimates in Table 4.9 are given in 1983 dollars.

TABLE 4.9 COST AND ENERGY ESTIMATES

	Capital Cost ($000)	Replacement Cost ($000)	Replacement Period (yrs)	Annual Operating Cost ($000)	Annual Maintenance Cost ($000)	Average Hourly Energy Savings (MMBtu)	Incremental Power for Electric Equipment (kW)
Steel							
Paper Hx-ZrO$_2$	102.9	61.7	5	10.4	20.6	7.02	5.5
Paper Hx-SiC	134.6	80.8	4	10.4	26.9	7.01	5.5
Woven Hx-ZrO$_2$	132.1	79.3	5	10.4	26.4	7.02	5.9
Woven Hx-SiC	176.8	106.1	4	10.4	35.4	7.01	5.9
SiC Compliant Seal Tube-in-Shell	250.9	150.5	5	10.4	50.2	7.02	5.9
SiC Spring-loaded Tube-in-Shell	263.2	157.9	3	10.4	52.6	7.21	27.6
SiC Tube-in-Tube	345.3	207.2	3	10.4	69.1	7.16	9.4
Heat Pipe	133.9	80.3	4	10.4	26.8	7.24	6.5
Aluminum							
SiC Compliant Seal Tube-in-Shell	338.2	202.9	1	10.4	67.6	7.90	19.3
Alumina-Chromia Helical	273.8	164.3	1	10.4	54.8	7.61	43.6
Paper Finned Plate Hx-ZrO$_2$	165.9	99.5	1	10.4	33.2	7.90	15.0
Woven Tube Hx-ZrO$_2$	205.3	123.2	1	10.4	41.1	7.90	14.8
Glass							
Alumina-Chromia Helical	1995.7	1197.4	1	10.4	399.1	123	490
Paper Finned Plate Hx-Al$_2$O$_3$	359.6	215.8	1/3	10.4	71.9	123	173
Woven Tube Hx-Al$_2$O$_3$	767.3	406.4	1	10.4	153.5	123	189

The replacement periods were conservatively assigned based on the relative corrosive resistance of the recuperator's materials of construction. In some cases, as in the aluminum application, all recuperators are assigned a conservative one year replacement period because lifetime is so uncertain. The basis for these assignments was discussed in Section 4.1.1. Assignment of a lifetime to the woven and paper concepts is difficult because so little is known about the behavior of the woven fibers and the necessary coatings. Therefore, these concepts were assigned the most conservative lifetimes, especially for the steel soaking application. The paper finned plate concept in the glass melting application was assigned the shortest replacement time of 1/3 of a year.

The energy savings in Table 4.9 reflect the average hourly savings while the facility is operating. This estimate includes typical temperature cycles for each application. This figure multiplied by the number of operating hours per year yields average annual energy savings. For all three applications, it was assumed that the facility is available 75% of the year (6570 hours). Somewhat similarly, the incremental power column in Table 4.9 gives the average increase in electrical demand required to support the recuperator mass flow rates. This figure multiplied by the number of operating hours per year yields the number of additional kilowatt hours used by the facility each year.

Energy savings are valued at $5/MM Btu for this economic analysis. In 1983 dollars this is quite conservative for natural gas, which currently averages nearly $9/MM Btu for industrial customers. The $5 figure is also conservative for residual fuel oil. One could justify using a higher value for fuel savings; however, the lower value takes into account regional differences in fuel costs without disturbing comparisons among the recuperators. Using a higher value would change the absolute but not the relative results of this economic analysis. Similarly, if the value of the fuel savings were escalated each year by some forecast inflation rate, the relative results of this analysis would be unchanged.

The cost of electricity was assumed to be $0.05/kWh. This is representative of the national average price paid by industrial customers.

Variations in this cost would have only a minor effect on the economic results of the economic analysis because annual electrical costs are small in comparison to other annual operating and maintenance costs.

As will be explained in Subsection 4.3.2, private enterprise would evaluate the economic attractiveness of an investment in recuperators on the basis of the recuperator's effect on the company's after-tax cash flow. For this reason, one must assume a tax situation for potential customers for these recuperators. Because the potential market for these recuperators is dominated by large companies, it was assumed that the user would be in a 46% tax bracket. This is the maximum federal income tax rate for corporations. It was further assumed that the recuperator would be eligible for a 10% investment tax credit and that it would be depreciated on a straight line basis. As with the value assigned to energy savings, variations in these assumptions would change the absolute results but would have little affect on relative comparisons between recuperators.

It was assumed that the industries interested in these recuperators would value energy savings and the associated costs over the entire life of their investment. Given recuperators with tube replacement periods of up to five years, ten years was chosen as the time horizon for this economic analysis. To compare costs and benefits over a multiple year period, one must assign a time value to the money involved. A 12% cost of capital was chosen for this analysis because it is representative of mature industries.

4.3.2 Economic Analysis Methodology

When corporate investments are optional, they are normally made on the basis of the investment's impact on after tax cash flows over the life of the investment. Using the cost estimates from Table 4.9 and the economic assumptions described above, the after tax cash flow for the Paper Hx-ZrO_2 recuperator in the steel soaking pit application is shown in Table 4.10 for each of 10 years included in the economic analysis. Each line in Table 4.10 is described below.

TABLE 4.10 AFTER TAX CASH FLOW FOR PAPER FINNED PLATE ZrO_2 HX

	Year 1	Year 2-4	Year 5	Year 6	Year 7-10
Energy savings	$230.8	$230.8	$230.8	$230.8	$230.8
Operating costs	10.4	10.4	10.4	10.4	10.4
Electrical costs	1.78	1.78	1.78	1.78	1.78
Maintenance costs	20.6	20.6	20.6	20.6	20.6
Depreciation	16.46	16.46	16.46	16.46	16.46
Before tax income	181.56	181.56	181.56	181.56	181.56
Increased tax	(83.52)	(83.52)	(83.52)	(83.52)	(83.52)
Tax credit	6.17			6.17	
After tax income	104.21	98.04	98.04	98.04	98.04
Capital replacement			(61.7)		
Depreciation	16.46	16.46	16.46	16.46	16.46
After tax cash flow	120.67	114.50	52.80	120.67	114.50

Energy Savings: Product of the average hourly energy savings from Table 4.9, the value assigned to energy--$5/MMBtu, and the number of operating hours per year--6570 hours.

Operating Costs: Taken from Table 4.9.

Electrical Costs: Product of incremental power from Table 4.9, the cost of electricity--$0.05 per kWh, and the number of operating hours per year--6570 hours.

Maintenance Costs: Taken from Table 4.9.

Depreciation: Of the original capital cost of $102.9, $61.7 is replaced after five years. Annual depreciation is then equal to the original capital cost less the replacement cost spread over 10 years plus the replacement cost spread over five years [i.e., (102.9 - 61.7)/10 + 61.7/5 = $16.5)]. Depreciation is a noncash expense which reduces tax liability but requires no cash expenditure.

Before Tax Income: Energy savings less annual costs and depreciation.

Increased Tax: 46% of the increase in "Before Tax Income."

Tax Credit: 10% of the capital investment in the previous year. In year 1, it is 10% of the capital cost. In year 6, it is 10% of the replacement cost in year 5. If the replacement period is less than three years, there is no tax credit.

After Tax Income: "Before Tax Income" less "increased taxes" plus the "tax credit" (if any).

Capital Replacement: This is an after tax cash expense. It is not tax deductible the year it occurs. Instead, it is depreciated over the life of the replacement.

Depreciation: The same amount that was deducted before determining tax liability is now added back because it was not a cash expense that year.

Cash Flow: The sum of after tax income plus depreciation less replacement costs for the year (if any).

The after tax cash flows can be compared to the initial capital cost to rank the recuperators on the basis of their relative economics. Several methods are in use by industry. Perhaps the easiest method is the simple payback. Payback is determined by calculating the amount of time it takes for the sum of the after tax cash flows to equal the magnitude of the initial cost. For example the paper Hx-ZrO_2 recuperator designed for a steel soaking pit initially cost $102.900. The first year's cash flow for this recuperator was $120,670 and the second year's cash flow was $114,500. So, the payback is one year plus (102,900 - 120,670)/114,500 years or about 0.9 years. The payback for each recuperator is listed in Table 4.11 for all three applications.

Because payback indicates how rapidly the initial capital cost is recovered, it gives a good measure of an investment's liquidity. While frequently used in industry, payback is normally used in conjunction with more sophisticated evaluation techniques. Payback is seldom used alone because, as the sole measure of an investment, it has two serious drawbacks. First, payback disregards all events subsequent to the payback period. For example, two investments each equal to $100 would each have a one year payback if they returned $100 the first year. However, if one of

TABLE 4.11 PAYBACK, RATE OF RETURN, AND NET PRESENT VALUE FOR BASE CASE CONCEPTS

	Payback (yr)	Rate of Return (%)	Net Present Value ($000)
Steel Application			
Paper Hx-ZrO_2	0.9	34	518
Heat pipe	1.0	31	484
Woven Tube Hx-ZrO_2	1.1	30	475
Lower Pressure Drop Case (4 psi gas/15 psi air vs. 8 psi gas/30 psi air)	1.7	23	332
Higher Fouling Factor Case (0.05 vs. 0.0)	2.0	21	284
Paper Finned Plate Hx-SiC	1.1	30	447
Woven Tube Hx-SiC	1.5	26	377
SiC Compliant Seal Tube-in-Shell	2.0	21	311
SiC Ball in Socket Joint Tube-in-Shell	3.6	17	197
SiC Tube-in-Tube	4.6	13	69
Aluminum Application			
Paper Finned Plate Hx-ZrO_2	2.6	21	192
Woven Tube Hx-ZrO_2	4.4	15	61
Alumina Chromia Helical Concept	10.0	0	-226
SiC Compliant Seal Tube-in-Shell	10.0	0	-388
Glass Application			
Paper Finned Plate Hx-Al_2O_3	0.2	56	9,507
Woven Tube Hx-Al_2O_3	0.4	46	9,738
Alumina Chromia Helical Concept	1.6	27	5,146

them returned $100 for ten years and the other returned $100 for only five years, the payback criterion alone would not distinguish between the two options. The 10 year investment is obviously superior.

The second drawback to using payback alone is that it does not consider the timing of the cash flow benefits. Again, looking at two $100 investments, suppose one returned $10 in year 1 and $90 in year 2, while the other returned $90 in year 1 and $10 in year 2. Both options have a two year payback, but option 2 is clearly superior because more interest could be earned during the second year on the increased first year cash flow.

To overcome the drawbacks associated with payback, industry uses discounted cash flow techniques such as an internal rate of return on net present value. These techniques take into account not only the magnitude but the timing of the cash flows over the entire life of an investment. The internal rate of return technique is not appropriate to investments in recuperators because the rates of return are too high and implicit in this technique is the assumption that the annual cash flows can be invested at the internal rate of return. The external rate of return circumvents this problem by reinvesting the annual cash flows at the company's cost of capital, 12% for this analysis. The external rate of return for each recuperator is shown in Table 4.11 along with the simple paybacks. Using the rate of return to rank the recuperators would alter the payback ranking.

Table 4.11 also lists the net present value (NPV) of each recuperator. The NPV is calculated by discounting all future benefits to the present at the cost of capital and then subtracting the initial capital cost. For example, the first year's cash flow is discounted by a factor of 1.12 (for a 12% cost of capital), the second year by a factor of 1.12 squared, etc. In this particular case the NPV ranking is not identical to the external rate of return ranking. A net present value of less than zero for this analysis would indicate that the recuperator does not achieve a 12% return over its lifetime.

4.3.3 Sensitivity of Economic Results to Input Parameters

The sensitivity of the new present value for the paper zirconium oxide heat exchanger in the steel application is shown in Figure 4.4 for varying cost estimates and energy savings. The Base Case conditions are given in Table 4.9. They are a capital cost of $102,900, an annual operating cost of $10,400, a five year replacement period and annual hourly energy savings of 7.02 million Btu valued at $5/MM Btu. To read Figure 4.4, enter with the percent deviation from the base condition on the horizontal axis, go to the line associated with the changed condition, and read the new net percent value on the vertical axis (e.g., if capital costs increase by 50%, net present value decreases from $518,000 to $480,000).

Considering that a new present value greater than zero is acceptable, the paper zirconium oxide heat exchanger appears to be acceptable for a 12% cost of capital, even in face of adverse changes in cost estimates, energy savings and the replacement period. It is most sensitive to revisions in the value of energy savings. This value can change by varying the average hourly savings in Btu, the value assigned per Btu to energy savings, or the number of operating hours per year. The sensitivity to revisions in the estimated replacement period is very high if the period decreases to one year. The rapid change between two years and one year can be attributed primarily to the different tax treatment accorded investments that last one year or less. Net present value is much more sensitive to capital costs than it is to annual operating costs. Although Figure 4.4 is based on the paper zirconium oxide heat exchanger in the steel soaking pit application, the sensitivity for other designs in other applications would be similar.

4.3.4 Variation in Economic Parameters with Variation in Selected Design Assumptions

Although the variation in economic parameters with economic inputs has been discussed generally, there are several important specific design assumptions which can be used to demonstrate the singular and combined effects of varied inputs on economic results. These are the replacement period which has a singular direct effect on the economics, and the maximum

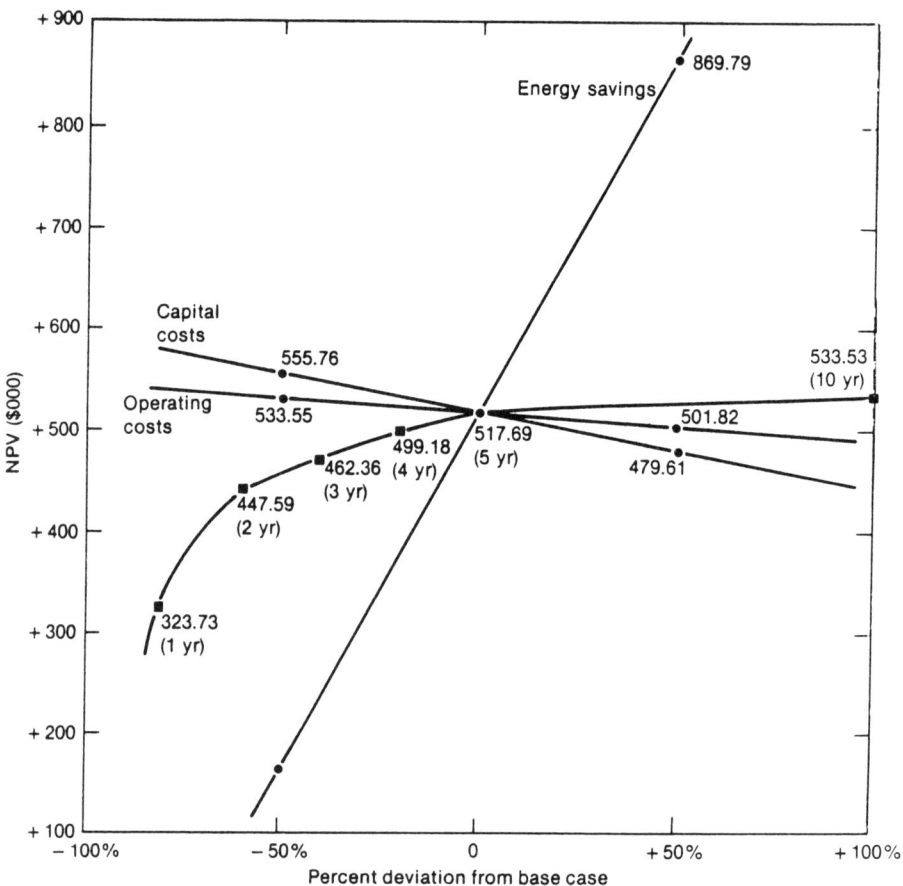

Figure 4.4 Net present value ($000) vs deviation from base case (%) for paper finned plate with ZrO_2 for steel application.

Evaluations   115

pressure drop and fouling factor which have a combined effect of capital cost, replacement cost, annual maintenance cost and incremental electric power on economics.

The replacement periods chosen are very conservative and may not be indicative of how long the given recuperators will last except in relation to one another. This is particularly true for the aluminum application where significant corrosion data do not exist (see Section 4.1) and a one year replacement period was assumed for all concepts. Therefore, a base case evaluation of the effect of a one to three year lifetime on economic parameters for the recuperators in the aluminum application was performed (see Table 4.12). These results indicate that if the paper finned plate unit had to be replaced on a yearly basis, it would still be more attractive than the SiC compliant seal tube-in-shell and alumina chromia helical units with replacement of three years. The woven tube unit with a replacement period of one year would be slightly worse than the other emerging units with replacement periods of three years.

The maximum pressure drops taken from the SiC compliant seal tube-in-shell case may vary because of the limited availability of specific fan powers or limitations in the industrial process to which the recuperator is to be retrofitted. The woven tube Hx-ZrO_2 in the steel application was designed with lower and higher maximum pressure drops in order to evaluate the effects of pressure drop. Details are included in Appendix F. The lower pressure drop case of 4 psi on the gas side and 15 psi on the air side compared to the base case of 8 psi on the gas side and 30 psi on the air side was further evaluated. The incremental fan power for the lower pressure case is 4.2 kW compared to 5.8 kW for the base case. The initial capital cost is $206.3K for the lower pressure case compared to $176.9K for the base case. Other economic values are included in Table 4.9 and in Appendix F where they are different for the lower pressure case. When economically evaluated, this results in a 1.7 year payback for the lower pressure case compared to a 1.1 year payback for the base case (see Table 4.11). This payback is still better than the paybacks for the state-of-the-art and emerging concepts equal to or greater than 2 years.

TABLE 4.12. VARIATION IN PAYBACK, RATE OF RETURN, AND NET PRESENT VALUE WITH CHANGE IN LIFETIME FOR THE ALUMINUM APPLICATIONS

	1 year	2 years	3 years
Paper Finned Plate Hx-ZrO$_2$			
Payback	2.6 yr	1.6 yr	1.2 yr
NPV	192K	392K	448K
Rate	21%	26%	28%
Woven Tube Hx-ZrO$_2$			
Payback	4.4 yr	2.3 yr	1.5 yr
NPV	61K	308K	377K
Rate	15%	23%	24%
Alumina Chromia Helical Concept			
Payback	>10 yr	4.5 yr	3.3 yr
NPV	-226K	103K	259K
Rate	<0%	15%	19%
SiC Compliant Seal Tube-in-Shell			
Payback	>10 yr	6.4 yr	3.6 yr
NPV	-388K	18.6K	243K
Rate	<0%	12%	17%

A zero fouling factor was assumed as the base case for all EG&G units based on a zero fouling factor in the SiC compliant seal tube-in-shell design. The effects on cost for a sample case (woven tube Hx-SiC in steel soaking pit application) are included in Appendix G. For a worst case fouling factor of 0.05, an initial capital cost of $234K was calculated compared to the $176.9K capital cost for the zero fouling factor base case. The incremental fan power for the 0.05 fouling factor case is 7.0 kW compared to 5.8 kW for the base case. Other cost values are included in Table 4.9 and Appendix G.

When economically evaluated, this results in a 2.0 year payback for the high fouling factor case compared to a 1.1 year payback for the base case. This is comparable to the paybacks obtained for the state-of-the-art and emerging concepts. The other EG&G concepts would be affected similarly by the addition of higher fouling factors particularly the paper finned plate concept with its higher probability of fouling (see Section 4.2.1). Depending on what fouling factor is assumed (this is not known for the waste streams of interest), the paper finned plate concept would probably have equivalent paybacks to the woven tube concept for the steel soaking pit application but better paybacks than the state-of-the-art and emerging concepts. With the higher fouling factors, the paper finned plate concept would probably have worse paybacks than the woven tube concept for the aluminum application and glass application but would still be better than the state-of-the-art and emerging concepts.

4.3.5 Potential Market

The potential for energy savings in the targeted applications for the steel, aluminum and glass industries has been documented in several reports dating from 1977.[9,10,11] Most of the marketing surveys performed in support of these reports have concentrated on characterizing the application and on estimating the total number of furnaces employed in each application. None of the references addresses the probability of actually penetrating the potential market with state-of-the-art ceramic recuperators. The following table is based on information in References 10 and 11.

Application	Number of Units	Annual Energy Savings (10^{12} Btu)	Percent Recouped
Steel Soaking	2275-2400	152-200	85-95
Aluminum	800	49-87	5
Glass	800-1000	180-260	10

On the basis of the percentage of units already recouped, the aluminum and glass industries appear to be good candidates for ceramic recuperators. However, in each of these industries, there are other considerations. Although only 10% of the glass furnaces are recouped, the other 90% use reputedly 70% efficient regenerators. Reference 11 estimates that only the 10% without regenerators would be serious candidates for ceramic recuperators. In the aluminum industry, the corrosive nature of the exhaust gases is a detriment to the use of recuperators.

Among the target applications, the steel soaking pit is the largest potential market, but as shown in the table, approximately 90% of these units are already recouped, but with mostly 40-50% effectiveness recuperators. The future of soaking pits, because they are being bypassed by new continuous casting operations and the industry is experiencing a severe economic slump, is also very uncertain. To penetrate this market, a new recuperator concept must be able to demonstrate that the incremental savings relative to the existing recuperator are sufficient to justify its cost. Such a comparison would require more detailed information about site specific existing recuperators than is available in the references.

4.3.6 Conclusions

In the steel, aluminum, and glass industries both EG&G recuperator designs are better or equal to emerging recuperators based on a best engineering judgment evaluation and an economic analysis.

In general, the economics can be most significantly improved by not having to replace a recuperator unit more frequently than every two years

and by increasing energy savings. A decrease in capital cost also has some
effect but not as significant. For example, a 25% improvement in energy
savings would have to be counteracted by a more than 100% increase in
capital cost for it to have a negative impact on the economics. Since the
costs were not developed independently and the scaled systems have
different loads on the ceramic portion of their systems, it would be
impossible to conclude anything more from the economic analysis.

The greatest potential market for improved ceramic recuperators is the
aluminum remelt furnace. Ten percent of the glass melting furnaces could
possibly use an improved replacement recuperator. The other 90% of the
furnaces would probably have to be totally replaced by a recuperative type
of furnace instead of new regenerative furnace. In the steel industry,
improved ceramic recuperators could replace the current recuperators in 90%
of the furnaces if use of steel soaking pits continues.

4.4 Summary of Evaluation

A consensus of the previously discussed evaluations follows. This
concensus is based on the evaluators' best engineering judgment regarding
durability, fouling, and fabricability, and on previously cited numbers in
the case of thermal performance and economics.

Many of the state-of-the-art and emerging recuperator concepts are not
usable as they are currently constructed in the selected waste heat streams
mainly because of durability of the material of construction (see
Section 4.1) and sometimes because of fouling and plugging of the small
passages (see Section 4.2). For this reason, only the concepts perceived
to have the potential to reasonably perform and survive in the harsh,
high-temperature environments of the three waste heat streams were
evaluated. It should be noted that slight modifications (especially in
material of construction) to the existing state-of-the-art and emerging
designs might have made these designs more amenable to use in the selected
waste heat streams, but these improvements will not be addressed here.

This preliminary elimination resulted in the following state-of-the-art and emerging concepts/applications being evaluated.

Concept	Application/Furnace Type
SiC Tube in Tube	Steel Soaking Pit
SiC Heat Pipe	Steel Soaking Pit
SiC Compliant Seal Tube in Shell	Steel Soaking Pit and Aluminum Remelt
SiC Ball and Socket Joint Tube in Shell	Steel Soaking Pit
Alumina Chromia Helical	Aluminum Remelt and Glass Melting

Since the steel soaking pit application has the least severe environment, four of the state-of-the-art and emerging concepts were relatively compared. With the information from the steel soaking pit evaluation and the realization that the aluminum remelt furnace environment is much more severe, only two representative state-of-the-art and emerging designs were relatively compared. Only one design was appropriate for the glass melting application.

The ranking of the above state-of-the-art and emerging concepts/applications was performed by relatively evaluating each concept/application based on its ability or inability to meet the design/functional/cost requirements (see Section 2) of durability (Section 4.2) performance (Section 4.3) and costs. These results were then compared with the economic results (see Section 4.4). It should be noted that each "ranking list" is in descending order of ability to meet the requirements; i.e. from best to worst. For steel soaking pit furnace applications, the ranking which resulted from this comparative evaluation is:

1. SiC compliant seal tube-in-shell concept
2. SiC tube-in-tube concept
3. SiC ball and socket joint tube-in-shell
4. SiC heat pipe concept.

The ranking which resulted from the economic evaluation for the steel soaking pit application was:

1. SiC heat pipe concept
2. SiC compliant seal tube-in-shell concept
3. SiC ball and socket joint tube-in-shell concept
4. SiC tube-in-tube concept.

For the aluminum remelt applications, using the comparative evaluation the ranking is:

1. SiC compliant seal tube-in-shell
2. Alumina Chromia Helical concept.

The ranking which resulted from the economic evaluation for the aluminum remelt application was:

1. Alumina Chromia Helical concept
2. SiC compliant seal tube-in-shell concept.

For the glass melting furnace a relative ranking of emerging concepts was not done since only one emerging concept was perceived as being appropriate for the extremely corrosive/fouling environments: the Alumina Chromia Helical concept.

These results were then compared by identifying areas of potential advantages and improvements with the EG&G concepts:

Woven Tube	
W/Al_2O_3	Glass Melting
W/SiC	Steel Soaking Pit
W/stabilized ZrO_2	Steel Soaking Pit and Aluminum Remelt
Paper Finned Plate	
W/Al_2O_3	Glass Melting
W/SiC	Steel Soaking Pit
W/stabilized ZrO_2	Steel Soaking Pit and Aluminum Remelt

4.4.1 Concept Evaluation/Comparison for the Steel Soaking Pit Furnace Applications

The EG&G concepts applied to a steel soaking pit furnace include the woven heat exchanger fabricated from stabilized zirconia (ZrO_2) or silicon carbide (SiC) and the paper heat exchanger fabricated from zirconia (ZrO_2) or silicon carbide (SiC).

The woven tube and paper finned plate heat exchangers have the potential for the highest costs because of the necessity for extra fabricating steps in making the thread for the woven tube, and in making the paper for the paper finned plate. Actual costs for fabrication are not available since these are not developed processes. Based on engineering judgment, the necessity for less material and relative ease of fabrication should outweigh the cost of extra fabricating steps and these two concepts should have comparable costs to the best emerging concepts, especially the paper finned plate concept which has the least material requirements. The woven tube concept using zirconia was judged to be comparable in terms of durability and to have potential for being better if further developmental information confirms the assumptions. The paper finned plate concept was judged to be good in durability because it is designed to be replaceable especially if significant fouling occurs and it cannot take the extra load. This problem of extra load with fouling may also be a problem with the woven tube concept. The woven tube concept was judged to be comparable to the best emerging concepts in terms of performance. The paper finned plate concept was judged to be slightly less comparable in thermal performance because of the higher probability of severe fouling.

Overall, both the woven heat exchanger and the paper heat exchanger fabricated from ZrO_2 ceramic fibers are judged comparable with the best of the emerging technology concepts, the SiC compliant seal tube-in-shell design and the SiC tube-in-tube concept. The woven tube heat exchanger with ZrO_2 appears to have the most potential with the paper heat exchanger fabricated from ZrO_2 not far behind.

Evaluations   123

When the concepts are compared purely on an economic basis (this analysis entailed a numerical consideration of thermal performance, materials costs and durability as discussed in the previous sections), the paper and the woven concepts are better than the best of the emerging technologies, the ceramic heat pipe and the SiC compliant seal tube-in-shell designs. The paper zirconia finned plate concept rates better than the woven zirconia tube concept until the probability of higher fouling in the paper finned plate design is considered in which case the two designs are probably comparable.

4.4.2 Concept Evaluation/Comparison for the Aluminum Remelting Furnace Applications

The EG&G concepts for application on an aluminum remelt furnace are the woven heat exchanger and the paper heat exchanger made from ZrO_2.

Both of the EG&G concepts were judged to be higher than either of the emerging concepts in the area of durability mainly because of the material and the threaded construction and both could be improved with further development. In addition, both concepts were comparable in terms of material costs and fabrication costs for the same reasons as the steel soaking pit application. The woven heat exchanger fabricated from zirconia was judged overall as high (in terms of performance) as the designs using the helical counterflow concept and compliant seal tube-in-shell concept and has potential for being better with further development. Both EG&G concepts, in particular the paper ZrO_2 finned plate concept, also rated higher than either of the emerging concepts with respect to economics, but again, the probability of severe fouling makes the woven tube concept better than the paper finned plate concept.

4.4.3 Concept Evaluation/Comparison for the Glass Melt Furnace Applications

EG&G's woven and paper heat exchangers made from alumina (Al_2O_3) are compared to the alumina chromia helical concept. Both EG&G concepts were judged to have better thermal performance characteristics than the helical concept, but the fouling aspects were judged to not be as good. Overall, both concepts, in particular the woven heat exchanger, compared favorably with this emerging technology concept. Both concepts, in particular the paper finned plate, also ranked higher than the emerging technology in terms of economics, but again, the probability of severe fouling downgrades the paper finned plate concept severely so that the woven tube concept is definitely the best.

4.4.4 Conclusions of Evaluation/Comparisons

Both the EG&G woven and paper heat exchanger concepts were judged to comparable overall with the state-of-the-art systems and emerging technology concepts developed by others based on a best engineering judgment evaluation and on an economic analysis. The area where the paper finned plate concept represents an improvement (especially compared to the one other heat exchanger in the same category, the cross-flow cordierite finned plate design) is its construction with a potentially more durable material (this needs to be confirmed) and its disposability if fouling becomes a problem. (Fouling prone areas, such as at the leading edge where impingement occurs and where the temperature is such that condensation occurs, can be replaced more often than the fouling-free areas.) Other advantages include the potential to minimize module to module leakage due to the EG&G corner post design. Another advantage is the potential for the use of inexpensive forms for weaving the cloth and folding the paper or cloth to form the heat exchanger shape, compared to extrusion machines for fabrication of monolithic units or parts of the emerging units.

The area where the ceramic woven tube heat exchanger represents an improvement (especially compared to other heat exchangers in the same category of tube-in-shell) is its potential for higher durability. It also has potential for less leakage through joints since it does not have joints. The other advantage is that because of the thinness of the material, the structure has the potential to be lighter, use less material, and need less support structure.

5. Research and Development Needs

Although a comparison and evaluation of state-of-the-art and emerging designs with advanced concepts has been done in this report, a number of assumptions were made due to the lack of information. Generally, when trying to evaluate even the developed existing designs, there is a severe lack of information which can be used to predict fouling and corrosion. Other research and development needs are discussed in more detail in the following subsections.

5.1 Fouling

Very little information is available in the open literature regarding fouling build-up in the three environments being considered in this study. Since the deposition rates and composition are so process dependent, it will be necessary to perform tests in the different exhausts for candidate materials:

1. Foulant build-up rate in representative modules at several different temperatures in each furnace waste gas stream under consideration should be obtained. Thickness histories can be used to determine the pressure drops in the passages and, together with loss in thermal performance, can be used to determine the cleaning frequency.

2. The foulant layer must be characterized, both by material properties and thermal properties. Posttest analysis will be required.

3. Cleanability and tenacity of the products need to be established. In most applications, it is desirable to remove the deposits with a dry technique, e.g., air lance, rather than a wet process where sulphuric acid is generated.

5.2 Corrosion

Very little corrosion information is available which can be used to predict the durability of materials used in existing and proposed designs for the three waste heat streams. In order to obtain this information, specimens characteristic of the actual material supplied by the manufacturer should be tested for lifetimes in actual or closely simulated waste streams under stress to simulate mechanical loadings.

Samples should be tested until failure with detailed fractography to determine the mechanisms which cause failure. The factors or variables that affect corrosion must be tested separately to maintain experimental control and combined effects must be determined.

5.3 Economics

Generally, the input for the economic analysis was adequate for the state-of-the-art and emerging designs, but the input for the advanced conceptual designs is not as well founded. As the conceptual designs are developed, the economics will need to be reevaluated and updated to be sure they are still favorable.

5.4 Performance

Although predicted and actual performance did not vary much since all concepts were forced to have a given performance, the predicted performance will need to be reevaluated as the concepts are developed and test data are obtained. The ability of the unit to attain this performance, especially over the lifetime of the unit, will be the best indicator for judging the concept's performance.

5.5 Woven Tube Heat Exchanger Concept

If the woven tube heat exchanger concept is to be further developed, several assumptions and potential problem areas need to be investigated. First, the behavior of the ceramic fibers at high temperature should be determined. Depending on these results, leakage may need to be controlled by the use of coatings which would then need to be investigated. Secondly, it should be determined if flow-induced vibration is going to be a problem and, if it is, design solutions will need to be found. Finally, a method for making the tube sheet to duct transition needs to be found.

5.6 Paper Finned Plate Concept

If the paper finned plate concept is to be further developed, several assumptions and potential problem areas need to be investigated. The effects of the thin material on increased leakage will need to be investigated. The high temperature strength, fatigue, and creep will also have to be determined for the candidate ceramic paper materials. Optimization may also be necessary to minimize fouling after the fouling mechanisms are understood.

6. References

1. EG&G Idaho, <u>An Assessment of Ceramic Materials Technology for Heat Exchangers</u>, EGG-SE-6367, November 1983.

2. R. L. Webb and A. K. Kulkarni, "Heat Exchanger Needs for Recovering Waste Heat in the Glass Making Industry," DOE/ID/12225, February 1983.

3. W. J. Marner and R.L. Webb (ed.), "Workshop on an Assessment of Gas-Side Fouling in Fossil Fuel Exhaust Environments," DOE/ID-12138-1, July 1982.

4. H. Hennekeu, Pittsburgh Plate and Glass, private communication, September 23, 1983.

5. W. Staniar, <u>Plant Engineering Handbook</u>, 2nd edition, 1959, pp. 9-25 to 9-27.

6. M. Peters and K. Timmerhaus, <u>Plant Design and Economics for Chemical Engineers</u>, 2nd edition, 1968, p. 132.

7. Richard A. Gaggioli, and T. J. Petit, Use the Second Law, First, Chemtech, Vol. 7, No. 8, pp. 496-506.

8. M. G. Coombs, D. M. Kotchick, H. J. Strumpf, "High Temperature Ceramic Recuperator and Combustion Air Burner Programs," GRI-82/0015, Garrett-AiResearch, 1982.

9. M. Gerstner and R. Stake, <u>Survey of Potential Energy Savings Using Effectiveness Recuperators for Waste Heat Recovery From Industrial Flue Gases</u>, TID-28954, AiResearch Mfg. Co., October 1977.

10. M. Coombs, et al., <u>High Temperature Ceramic Recuperators and Combustion Air Burner Programs - Annual Report</u>, GRI-82/0015, April 1982.

11. V. J. Tennery, <u>Economic Applications Design Analysis, and Material Availability for Ceramic Heat Exchangers</u>, ORNL/TM-7580, January 1981.

Appendix A

Drawings and Descriptions of Materials Concepts

CRACK STOPPING ADDITIVES   MATERIALS

* ADD HIGH TEMP. VERSIONS OF 3M'S MICROSPHERES TO CERAMIC STOCK
* DEFORMATION OF PARTICLES DISSIPATES CRACK ENERGY
* MAXIMUM DISTANCE BETWEEN PARTICLES IS SOME FUNCTION OF CRITICAL FLAW SIZE

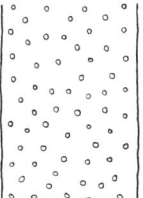 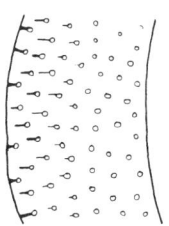

CRACKS STOP AT
MICROSPHERE/CERAMIC
INTERFACE

PIEZOELECTRIC CERAMICS   MATERIALS

* STRAINING CAUSES BUILD-UP OF ELECTRIC POTENTIAL OR CHARGE
* COULD BE USED AS AN NDE TOOL
* CHARGE BUILD-UP WOULD ATTRACT "PLUGGING" PARTICLES IN SUSPENSION
* STRAINING BEYOND CRACK ENERGY THRESHOLD CAUSES A CRACK THAT RELEASES A SPARK - SPARK INITIATES A "THERMIT WELDING" PROCESS AMONG "PLUGGING" PARTICLES WHICH FLOW INTO THE CRACK, BONDING TO THE WALLS AND SEALING THE CRACK

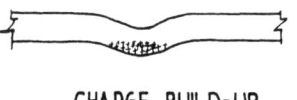

CHARGE BUILD-UP

ATTRACTION OF PLUGGING PARTICLES

PARTICLES PLUG CRACK

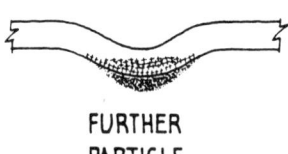

FURTHER PARTICLE BUILD-UP

OR

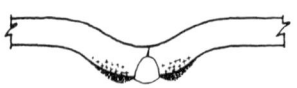

THERMIT WELD TRIGGERED BY SPARK

COMPOSITE CERAMICS   MATERIALS

* STRAINS CRACK MATRIX, BUT STOP (DISSIPATE ENERGY) AT FIBER/MATRIX INTERFACE
* MATRIX FUNCTIONS AS AN ENVIRONMENTALLY PROTECTIVE COVER FOR FIBERS: COULD BE SLIGHTLY PLIABLE AND NON-STRUCTURAL
* HIGH NUMBER OF SMALL DIAMETER, INDEPENDENT FIBERS ACT AS ACOUSTIC EMITTERS WHEN OVERSTRESSED, WARNING OF LOAD TRANSFER (WOOD FIBER ANALOGY) MAY PREVENT CATASTROPHIC FAILURE
* MIX HIGH AND MID-STRENGTH FIBERS: MID-STRENGTH FIBERS CRACK FIRST, INDICATE POTENTIAL OVERSTRESS LIMIT
* DEVELOP HEALING RESETTING MID-STRENGTH FIBERS

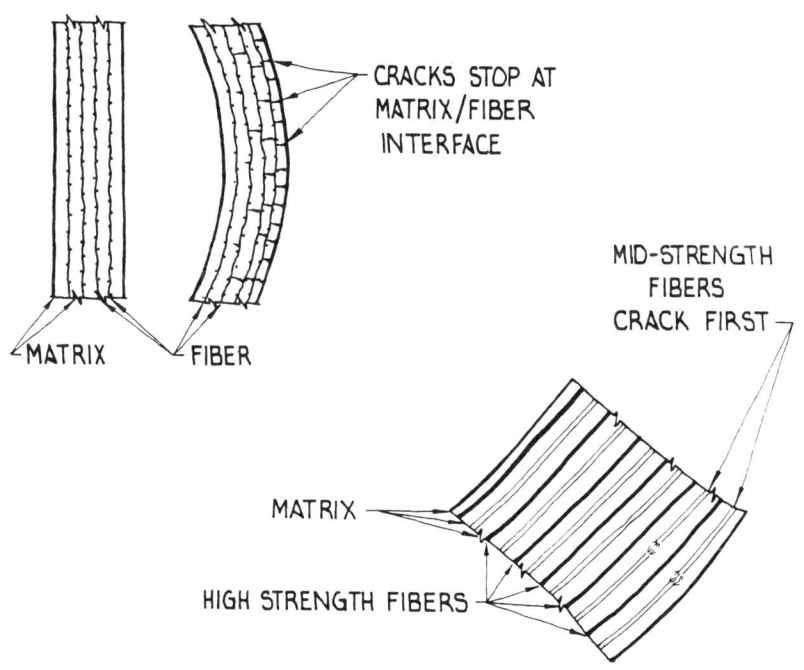

FLEXIBLE CERAMIC MATERIALS

* USE OF COMPOSITES TO ACHIEVE FLEXIBILITY
* USE OF CHEMICAL/MECHANICAL COMPOUNDING
* BENDABLE FIBERS WITH GEL-LIKE MATRIX

Appendix B

Drawings and Descriptions of Fabrication Concepts

WOVEN TUBE FABRICATION

* WEAVE TUBE FROM HIGH TEMPERATURE FABRIC/FIBERS
* GEOMETRY: ROUND, SQUARE, ELLIPTICAL, ETC.
* MINIMIZE LEAKAGE BY:
  - TIGHT WEAVE
  - AIR PRESSURE CONTROL TO ELIMINATE FLUE GAS/AIR ΔP
  - VISCOUS LIQUID (@OPERATING TEMPERATURE) RESIDES IN PORES DUE TO SURFACE TENSION... COATS FABRIC FIBERS TO PROVIDE ENVIRONMENTAL RESISTANCE... PERMEATED FABRIC IS LUBRICATED & CUSHIONED DURING FLEXING

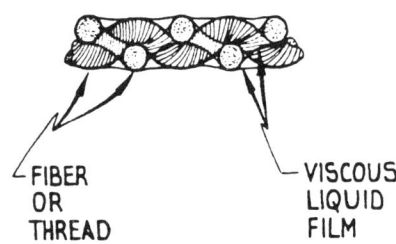

WRAPPED PAPER OR FABRIC FABRICATION

* CONSTANT WIDTH PAPER OR FABRIC FED IN ON BIAS, TWISTED TO CLOSE SEAM, THEN BONDED OR SEWN SHUT
* MULTI-LAYERED, OPPOSED BIASING FOR TORSION AND BEND RESISTANT TUBE
* ANY LENGTH AND SIZE POSSIBLE (WITHIN MATERIALS LIMITS)
* VISCOUS LIQUID COATING COULD SEAL TUBE AND OR LUBRICATE FIBERS

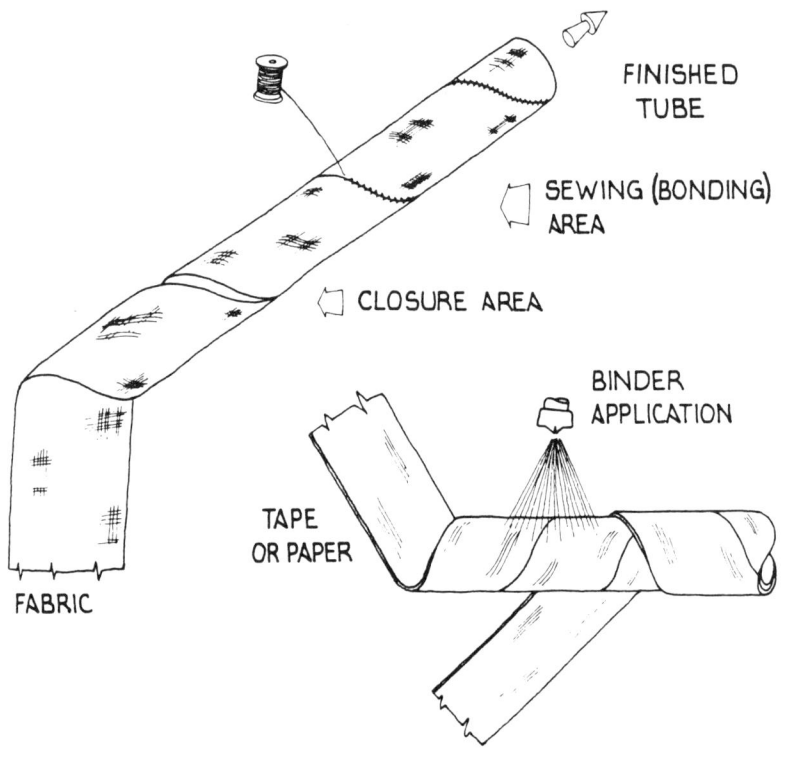

OPPOSED BIAS CONSTRUCTION

Appendix B: Fabrication Concepts   135

CERAMIC COATED FIBERS   FABRICATION

* FIBERS ARE COATED WITH CERAMIC
* FIBERS BURN OR DISSOLVE OUT, LEAVING BEHIND A HOLLOW SPACE
* HOLLOW FIBERS HAVE HIGH SURFACE AREA TO VOLUME RATIO
* IF FIBERS ARE FLEXIBLE, HIGH FLOWRATES MAY GENERATE VORTEX STREET OSCILLATION, AUGMENTING HEAT TRANSFER

1

BARE
FIBER

2

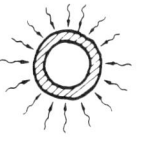

CERAMIC
COATING
APPLIED

3

FIRING
BURNS OUT
FIBER

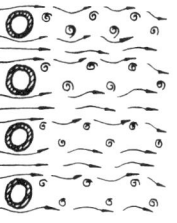

VORTEX STREET
GENERATION

COMPRESSION COATING   FABRICATION

* DEVELOP A COATING THAT FORMS A SKIN OVER THE CERAMIC PART
* SKIN IMPARTS A COMPRESSIVE LOADING ON PART WHILE MAINTAINING A TENSILE STRESS CONDITION IN ITSELF (SHRINK-WRAP ANALOGY)

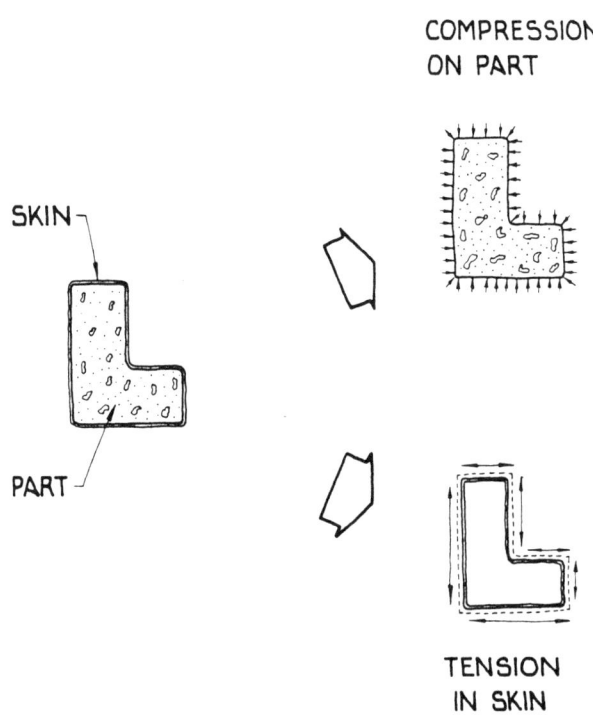

Appendix B: Fabrication Concepts   137

GRADED MATERIAL TUBES   FABRICATION

* TUBE ENDS HAVE DIFFERENT PROPERTIES THAN CENTER SECTION
* FORMED IN PLACE COMPLIANT SECTIONS RELIEVE STRAIN....
  POSITIONED BETWEEN RIGID ENDS

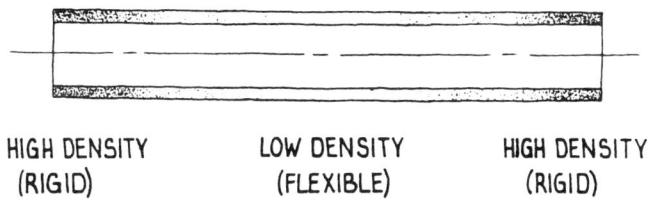

HIGH DENSITY   LOW DENSITY   HIGH DENSITY
(RIGID)   (FLEXIBLE)   (RIGID)

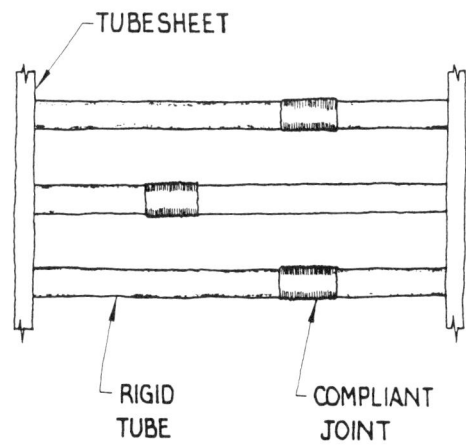

TUBESHEET

RIGID TUBE   COMPLIANT JOINT

WOUND TUBE FABRICATION

* WIND HIGH TEMP. FIBER LIKE ROCKET CASING: ORBITAL OR BROADGOODS
* SPRAY VISCOUS LIQUID OR OTHER HIGH TEMP. BINDER DURING WINDING FOR BONDING

VIBRATING FABRICATION FABRICATION

* VIBRATION ENHANCED SETTLING: MINIMIZES VOIDS AND CRACK INITIATION SITES
* SLIP CASTING - CONCRETE POURING ANALOGY
* EXTRUSION - VIBRATE RAW STOCK TO DENSE, HOMOGENEOUS CONDITION PRIOR TO EXTRUSION
* HIPPING - INERTIAL FORCES AID PRESSURE FORCES

Appendix B: Fabrication Concepts

COMPRESSED FIRING   FABRICATION

* FIRE GREEN STOCK IN A COMPRESSED CONDITION
* FLAWS MAY NOT APPEAR DURING FABRICATION

POWDER PRODUCTION   FABRICATION

* CAVITATING DEVICE FOR PRODUCING FINE POWDERS
* ACCELERATED EROSION: WIND, WATER, VERY HARD SOLIDS
* PARTICLE SEPARATION BY MICRON MESH SCREENS, SETTLING TANKS, CENTRIFUGE PROCESSES, AEROSOL EVAPORIZATION, ETC

PLIABLE CERAMIC   FABRICATION

* CERAMIC IS PLIABLE AT OPERATING TEMPERATURES
* STRAINS ARE RELIEVED IMMEDIATELY THROUGH DEFORMATION

Appendix C

Drawings and Descriptions of Component Concepts

HIGH TEMP. PAPER Hx   COMPONENT

* USE IT TILL IT FOULS, THEN THROW IT AWAY OR RECYCLE/RECLAIM
* PAPER CONSTRUCTION MEANS LOW MASS IMPLYING LOW COST AND RESISTANCE TO THERMAL SHOCK
* FABRICATION WITH CERAMIC GLUE/ADHESIVE - FIBERS IN BINDER
* UNITIZED MOLD PRODUCES MODULAR COMPONENTS
* HANGING FAN SHAPED Hx: STACKED STRIPS WITH SIDES ALTERNATELY BONDED TOGETHER - HANGING OR EXPANDED WHEN IN POSITION; FORMS BARRIER BETWEEN HOT AND COLD STREAMS
* PAPER FIBERS CAST WITH CEMENT BINDER - USE 'LOST WAX PROCESS' TO REMOVE MOLD CORE DURING FIRING

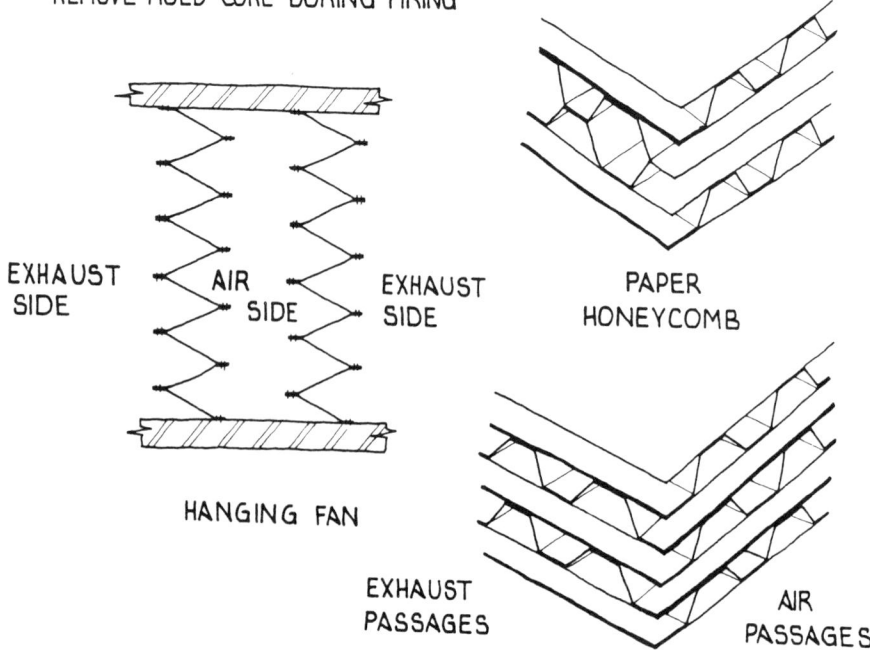

INFLATABLE/EXPANDABLE TUBESHEET COMPONENT

* PRELOADS TUBES IN COMPRESSION
* SANDWICH CONSTRUCTION - FILLING IS MATERIAL THAT AT OPERATING CONDITIONS EXPANDS SLIGHTLY (AS IN PHASE CHANGE) PRELOADING THE JOINT
* NECESSITATES FLEXIBLE CERAMIC TO FLEX AGAINST TUBES
* COULD INFLATE WITH AIR PRESSURE - MIGHT USE INFLATED VOID AS AIR DISTRIBUTION PLENUM

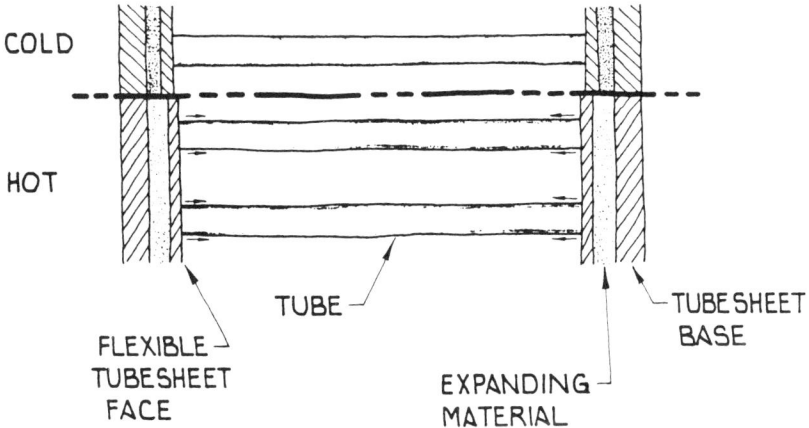

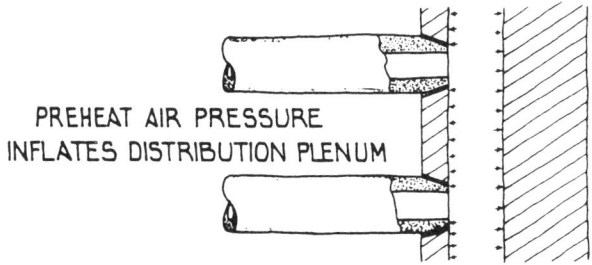

EXPENDABLE-EXTRUDABLE SEAL   COMPONENT

* PUMPED TO TUBE TUBESHEET INTERFACE, SEALS, AND VOLATIZES AS IT LEAKS INTO THE EXHAUST STREAM
* CONDENSES OUT AT TAIL END OF Hx WHERE IT IS COLLECTED AND REUSED: MIGHT BE USED TO BRING POLLUTANTS OUT OF EXHAUST
* MIGHT USE PROCESS FLUID: i.e.- ALUMINUM

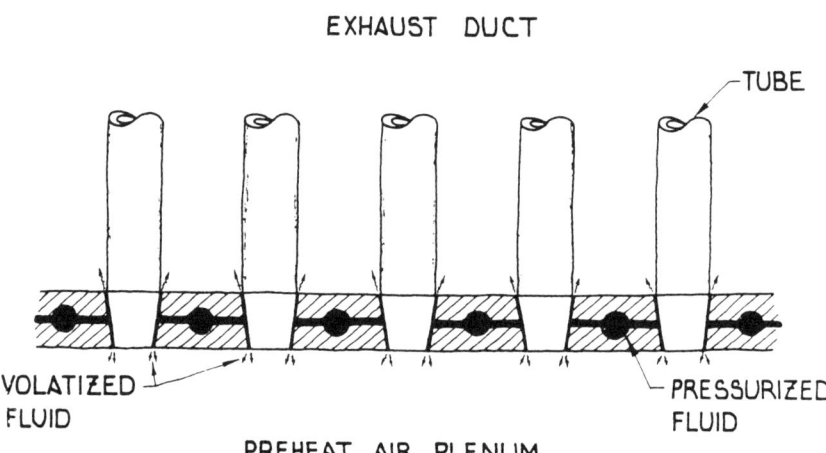

STACKED CYLINDERS W/ VISCOUS LIQUID COMPONENT

* FLAT DISCS OF INCREASING DIAMETER WITH SHOULDERS STACK BETWEEN TUBE AND TUBESHEET - VISCOUS LIQUID IN GAP
* TOLERANCE STACKUP ALLOWS FOR LARGER LATITUDE IN GROWTH AND MISALIGNMENT
* REQUIRED ALLOWANCE OF MOTION DEFINES NUMBER OF DISCS

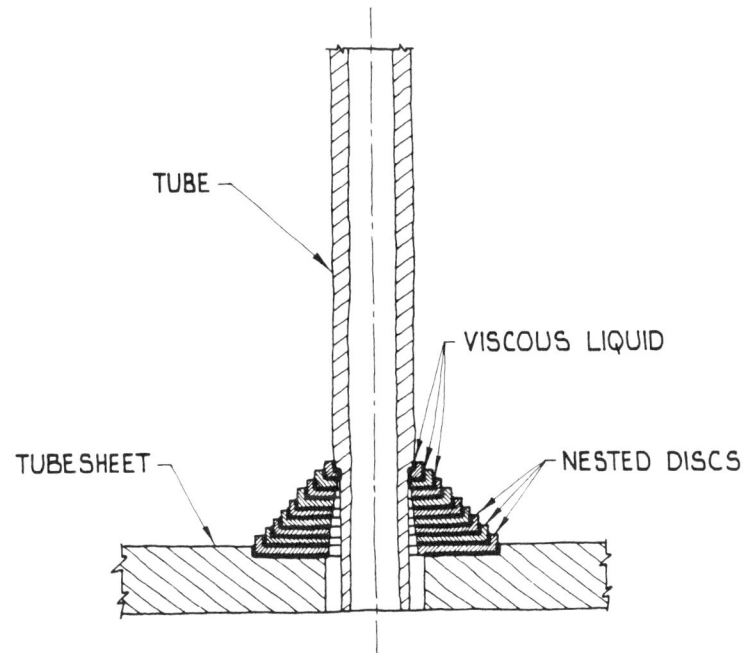

'MARCHING' REPLACEABLE Hx   COMPONENT

* FOULING PRODUCTS ARE GENERALLY (DURING DESIGN CONDITIONS) DEPOSITED AT CERTAIN TEMPERATURE RANGES FARTHER BACK IN THE Hx
* WITH MODULAR CONSTRUCTION AND EXPENDABLE/LIMITED LIFETIME COMPONENTS, NEW MODULES ARE INSERTED AT THE FRONT, FOUL PRONE OR DEGRADED MODULES ARE REMOVED AT THE BACK
* MODULAR CONSTRUCTION LIMITS OFF-DESIGN EXCURSION DAMAGE TO ONLY THOSE COMPONENTS EXPERIENCING DAMAGING CONDITIONS

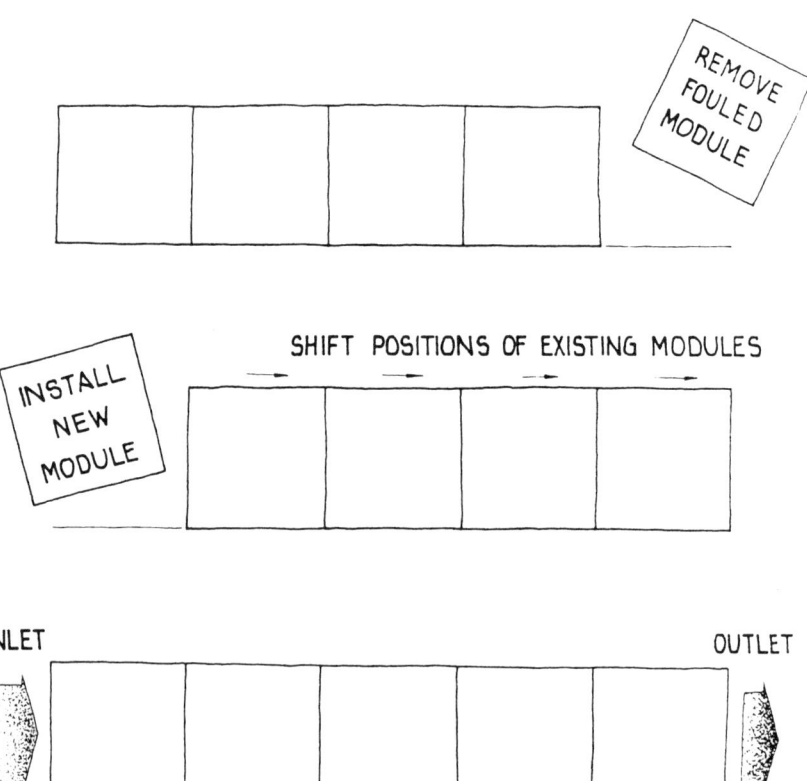

Appendix C: Component Concepts   145

ISOSTATIC JOINT COMPRESSION COMPONENT

* LOADING OF MECHANICAL SEALS IS BY AN INFLATABLE BAG PRESSURIZED WITH AIR; LEAKAGE IS PERMISSIBLE WITH AN ADEQUATELY SIZED PUMP
* LOAD IS TRANSMITTED FROM BAG THROUGH INSULATION TO Hx COMPONENT STRUCTURE/HOUSING
* ALTERNATIVELY, HIGH DENSITY SECTIONS (IN A LINEAR OR GRID PATTERN) COULD TRANSMIT LOAD WITH THE REMAINING AREA FILLED WITH LOW DENSITY INSULATION
* BAG IS OUTSIDE OF HIGH TEMPERATURE REGION, ALLOWING LOW COST MATERIAL USE
* WITH FLOW THROUGH BAG AREAS AND HIGHER TEMPERATURE MATERIALS, AIR COULD BE PREHEATED SLIGHTLY PRIOR TO ENTERING THE Hx

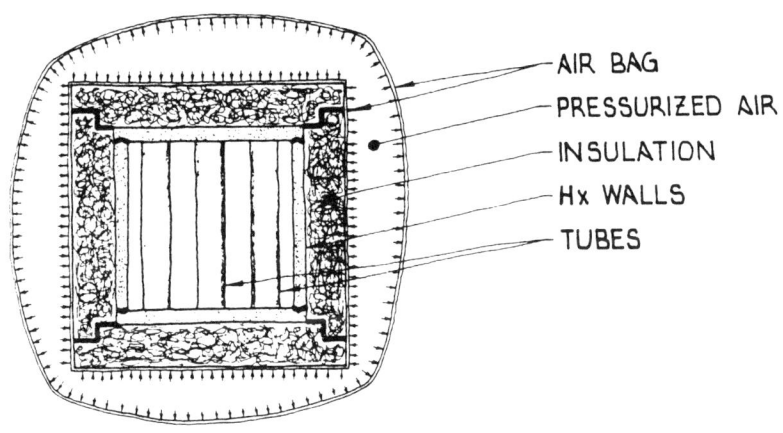

AIR BAG
PRESSURIZED AIR
INSULATION
Hx WALLS
TUBES

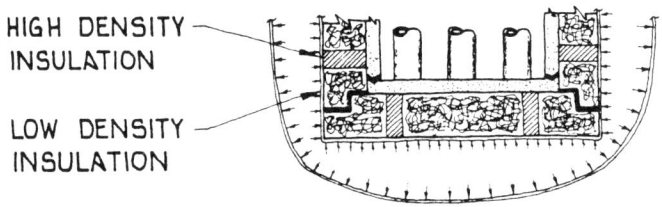

HIGH DENSITY INSULATION
LOW DENSITY INSULATION

LIQUID COVERED TUBESHEET   COMPONENT

* VISCOUS LIQUID FORMS SEAL BUT ALLOWS THERMAL EXPANSION
* LIQUID DAMPENS VIBRATION

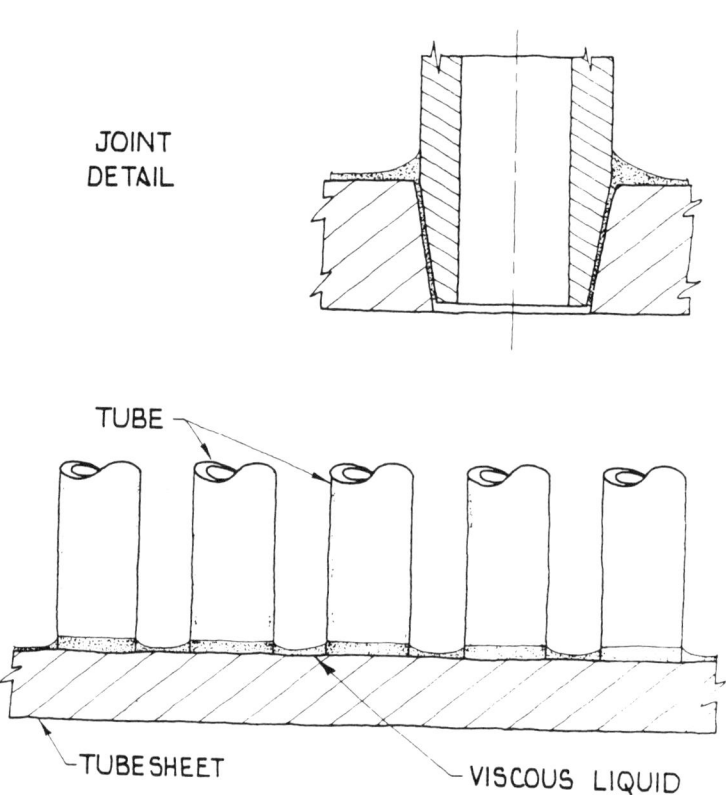

JOINT DETAIL

TUBE

TUBESHEET

VISCOUS LIQUID

ALLOY MIGRATION   COMPONENT

* HIGH CONCENTRATION OF ALLOYING ELEMENTS HAS HIGHER MELTING PT.
* ALLOYING ELEMENTS MIGRATE TO COLDER AREA RAISING M.P.
* AS ELEMENTS MIGRATE AWAY, HIGHER PURITY MATERIAL SOFTENS AND MAY MELT
* SOFT TUBE/TUBESHEET INTERFACE SEALS AND ABSORBS THERMAL GROWTH @ OPERATING TEMPERATURE

OSCILLATING BUBBLE PROCESS

* AIR PULSES CAUSE BUBBLE EXCURSION INTO EXHAUST STREAM
* RESONANT AIR PULSES CAUSE OSCILLATING EXHAUST FLOW DUE TO BUBBLE INFLATION IMPEDING EXHAUST FLOW TEMPORARILY UNTIL THE BUBBLES COLLAPSE. (COMPRESSION-RAREFACTION)
* PULSE COMBUSTION PROCESS (LIKE V-1 ROCKET) TO DRIVE AIR PULSES
* EXHAUST PULSES MAY BE DAMPED DURING PROCESS BUT ARE RESTORED BY OSCILLATING BUBBLES

OSCILLATING BUBBLE

Potential Advantages

May have small size.

Can pump fluid due to peristaltic motion.

Potential Areas of Concern

Bubble material (fabric or film) unobtainable at this time.

Need extremely small pores and bubbles.

Bubble material may be fouled by waste stream.

Appendix D

Descriptions of Process Concepts with Lists of Potential Advantages and Disadvantages

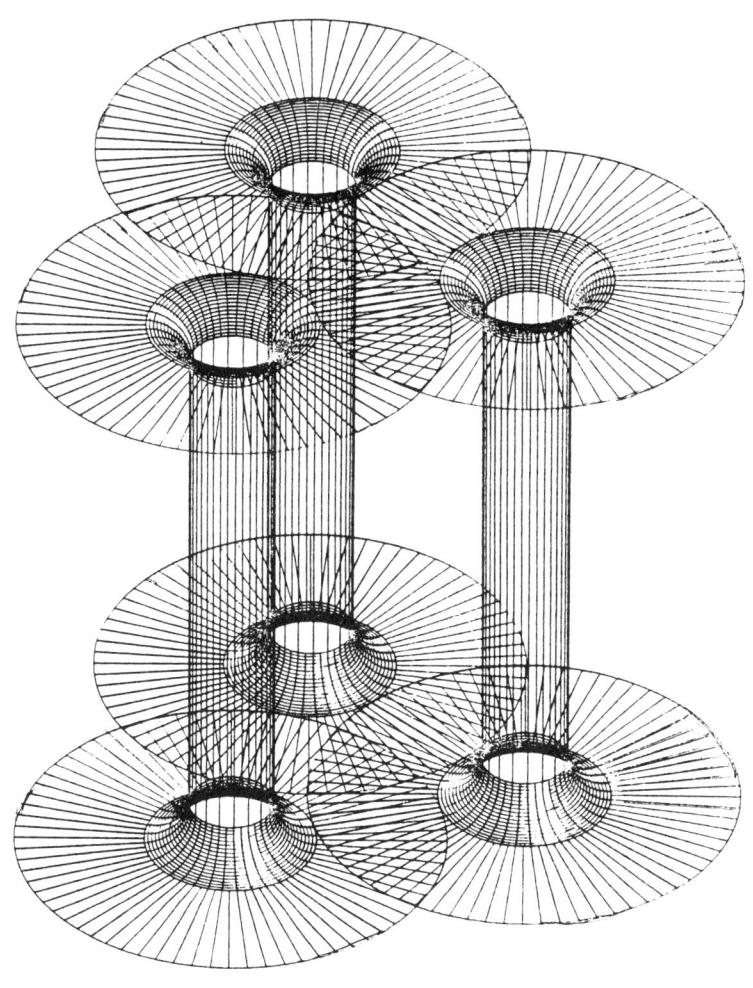

WOVEN TUBE/TUBESHEET

Process Concept

Potential Advantages	Potential Areas of Concern
Smooth transition into flow paths for both streams minimizes pressure drop.	Smooth transition into flow paths minimized heat transfer.
Low amount of material usage base on size (density) keeps costs down.	Possible leakage through uncoated weave; may dilute combustion air and interfere with heat transfer.
Fibers share loads; if one breaks not a catastrophic failure, just pinhole leaks; (redundant load paths).	Possible flow induced vibration in flexible tube.
Constant thickness junction between tube and tubesheet minimizes thermal stresses and strains.	Possible problem with tubesheet to duct transition.
Structure is somewhat more flexible (load shifting) than normal tube and shell H_x.	Possible loss of strength of fibers at temperature.
Many fabrication methods are possible based on a basic weaving technique; different material fibers can be interwoven for a fiber only composite; any shape is possible.	

Appendix D: Process Concepts 151

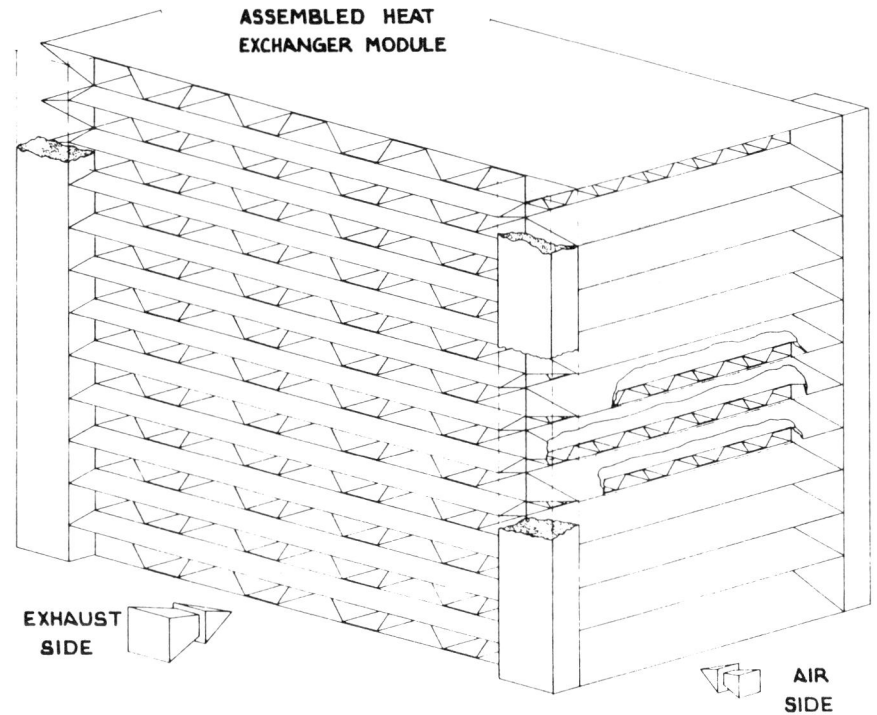

PAPER HONEYCOMB H_x

Process Concept

Potential Advantages	Potential Areas of Concern
Versatility - small modules may be stacked; easy to install; easy to repair; easy to dispose of; easily retrofitted to square or rectangular ducts; easily shipped and modules replaced.	Fouling of small passages. Possible leakage through paper, fabric, or mat (porosity?).
Fibers share loads; if one breaks, pinhole leaks instead of catastrophic failure (redundant load paths).	Bonding agent in paper may volatize and cause paper to lose strength.
Rigid lightweight structure.	Sealing between modules may be a problem.
Very thin materials - potentially less costly.	Relies totally on bonding technology.
Entrance and Exit venturies minimize pressure drop.	Entrance and Exit venturies minimize heat transfer.

PHASE CHANGE SHOT TOWER PROCESS

* UTILIZES RAINING BED TECHNOLOGY
* SHOT FALLS THROUGH EXHAUST STREAM, ABSORBS HEAT, AND MELTS
* LIQUID IS COLLECTED AND TRANSFERED TO NEXT TOWER
* LIQUID IS SPRAYED (THROUGH A NOZZLE) OR DISTRIBUTED (ROTATING DISC/CONE) WHERE IT FALLS THROUGH COMBUSTION AIR, PREHEATS IT. AND SOLIDIFYS INTO SHOT AS IT FALLS
* SHOT IS COLLECTED AND RETURNED TO EXHAUST STREAM INPUT

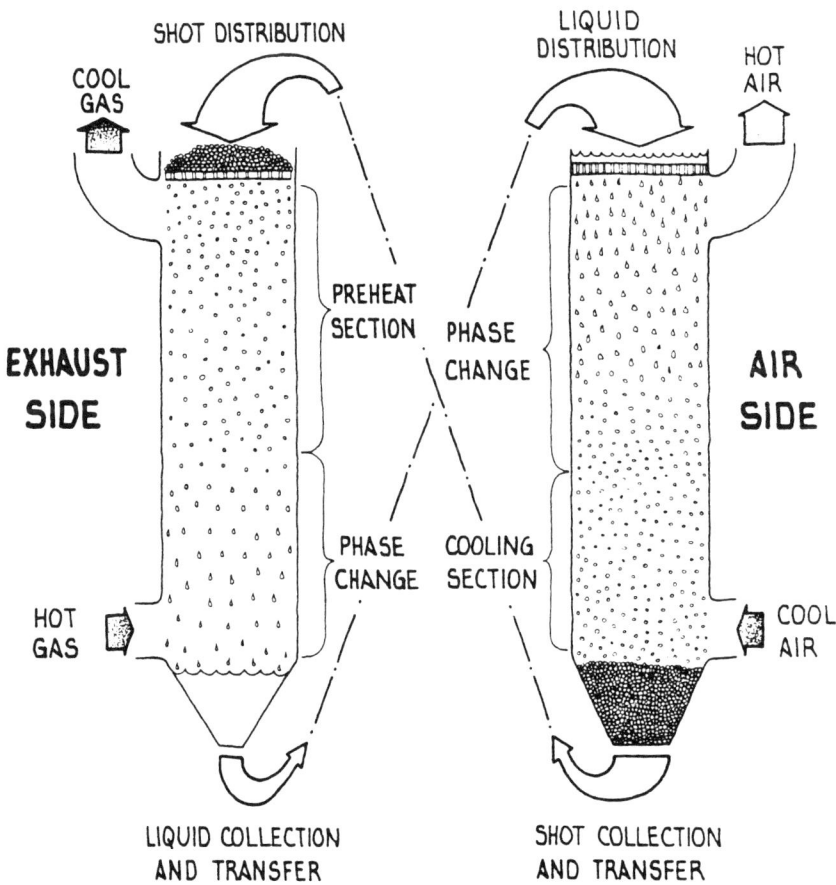

PHASE CHANGE SHOT TOWER

Potential Advantages

Uses latent and sensible heat recovery.

Smaller space than with just sensible heat recovery.

Liquifying action can be used to purify waste stream.

Lower pressure drop than packed fluidized bed.

Potential Areas of Concern

Material not available yet.

High temperature distribution and control system needed to handle liquid.

Narrow melting range required.

Not as compact as packed bed.

Appendix D: Process Concepts   155

COATED SPONGE   PROCESS

* IMPERVIOUS COATING OVER SHAPED CERAMIC SPONGE OR MASSED FIBERS: COATING SEALS SPONGE/FIBER SHAPE
* EXHAUST HEAT TRANSFERRED THROUGH COATING/AIR PREHEATED BY PASSING THROUGH SPONGE
* CLEAN FOULING PRODUCTS BY DISSOLVING OR MECHANICALLY REMOVING COATING, THEN RECOAT
* GEOMETRY: FLAT PLATE OF "X" THICKNESS WHICH IS PERFORATED AND POSITIONED NORMAL TO THE FLOW (CROSSFLOW Hx)
  COULD ALSO BE POSITIONED ON EDGE AS A PLATE/SURFACE Hx, BUT SHOULD USE HEAT TRANSFER ENHANCEMENT METHODS TO IMPROVE $\bar{h}_{surf}$

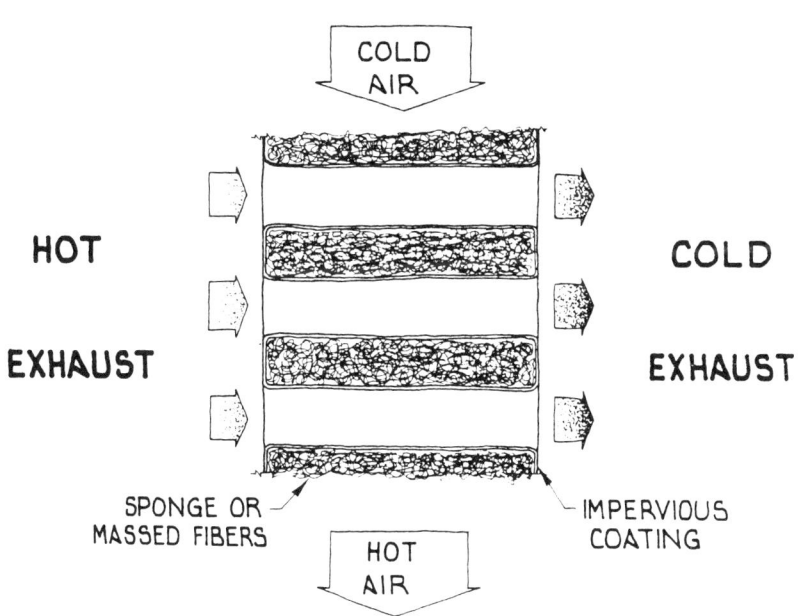

COATED SPONGE

Potential Advantages

Acts as filter (in exhaust stream).
High convective coefficient.
Fibers might be structural.
Small size.

Potential Areas of Concern

Long conduction path: Through coating and then along fibers.
Could be fouling site.
High pressure drop.

CROSSFLOW LIQUID BATH   PROCESS

* COMMON TUBES IN A CROSSFLOW CONFIGURATION ARE SURROUNDED BY A MOLTEN LIQUID
* HEAT TRANSFER TO LIQUID BY NATURAL CONVECTION
* ENHANCED/AUGMENTED HEAT TRANSFER METHODS NEEDED ON INTERIOR OF TUBES ONLY

CROSSFLOW (OR COUNTERFLOW) LIQUID BATH

Potential Advantages	Potential Areas of Concern
High thermal inertia damps out transients and thermal shock effects.	High thermal inertia.
	4 convective coefficients, 2 natural with liquid.
	Weight of passages and liquid.
	Large size.
	Large amount of liquid and materials.

TRI-CYCLE REGENERATOR PROCESS

*DELTOID PROCESS DIVISION: HEAT-COOL-CLEAN
*CLEANING CAN BE DONE EVERY CYCLE OR OTHER SPECIFIED PERIOD

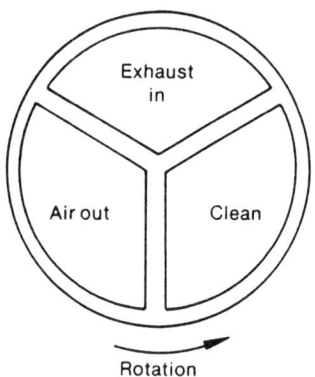

TRICYCLE REGENERATOR STAGING

Potential Advantages

High temperature cleaning can minimize thermal shock.
Small passages good for heat storage.
Can be relatively compact.
Existing designs can be modified or retrofit.
Staging with fouling filters possible.

Potential Areas of Concern

Cuts down on available porting space.
Low temperature cleaning produces large thermal shock.
Small passages foul easily.
Sealing.

GAS ENTRAINMENT IN LIQUID   PROCESS

* ENTRAINED GAS TRANSFERS HEAT TO LIQUID
* GAS BUBBLES COLLAPSE (OR SHRINK) AS THEY GIVE UP HEAT
* AIR BUBBLES EXPAND AS THEY ABSORB HEAT
* LIQUID CAN BE USED TO FILTER EXHAUST GAS - CLEAN FLUID WHEN COOL

GAS ENTRAINMENT IN LIQUID

Potential Advantages	Potential Areas of Concern
Large amount of heat.	Large liquid inventory (relative to gas flow) implies large amount of low grade heat.
Extraction.	
Can clean waste stream of particulates and such.	High temperature liquid mandates high temperature pumps, metering devices, and collectors.
	Very large size.

LENSES TO CONCENTRATE HEAT   PROCESS

- FRESNEL OR COMPOUND EYE APPROACH TO CONCENTRATE HEAT
- COULD BE AT PROCESS AREA IN FEEDSTOCK FILL ZONE AND/OR IN EXHAUST ZONE
- RADIATIVE HEAT TRANSFER AUGMENTS SENSIBLE HEAT EXCHANGE

LENSES TO CONCENTRATE HEAT

Potential Advantages

Recovers radiant energy.

Can concentrate energy.

Could be placed looking into stack after first turn.

Potential Areas of Concern

Large lenses required even with Fresnel type.

Might foul (get dirty).

Needs receiver to absorb heat.

SPIRALLY FLUTED TUBING HEAT PIPE   PROCESS

* SPIRAL FLUTES ACT AS WICK ELIMINATING SEPARATE PART
* SPIRAL FLUTES IMPART CENTRIFUGAL FORCE TO FLUID ADDING ENTRAINMENT RESISTANCE TO INTERIOR FLOW
* ENHANCED EXTERNAL HEAT TRANSFER COEFFICIENT DUE TO EXTERNAL FLUTING
* OPTIMIZE SPIRAL HELIX ANGLE, NUMBER OF FLUTES, AND FLUTE DEPTH
* COAT INSIDE OF HEAT PIPE TO PROTECT IT FROM WORKING FLUID
* COULD USE FABRIC REINFORCED CERAMIC COMPOSITE

CERAMIC HEAT PIPE

Potential Advantages

Redundant heat transfer paths heat pipe is not heat transfer limit.

Large heat fluxes in small cross-section.

Many shapes possible.

Gas loading minimizes thermal shock (variable conductance).

Composite construction may alleviate working fluid attack and environmental degradation.

Potential Areas of Concern

Current working fluids attack pipe walls.

Working fluids are hazardous substances.

Seals can be a problem (hermetic).

: LASL is actively involved here. Garrett has submitted a proposal to work with LASL.

Thermo Electron (TECO) is working in this area.

JET IMPINGEMENT   PROCESS

JET IMPINGEMENT (WITH CYCLONE)

Potential Advantages

Cyclonic action tends to filter waste stream.

Potential Areas of Concern

Holes might foul and plug.

High pressure losses (potential).

: Based on a low temperature idea from Thermal Systems Engineering (GRI sponsored).

TETRAHEDRON HX   PROCESS

TETRAHEDRON Hx

Potential Advantages

Thermal transients only in center plates.

Very crush resistant shape.

Compression loaded structure.

Routing of flow possible by orienting cells differently.

Turbulence generating structure for high Hx coefficients.

Can be slip cast; can be stamp molded; can be made from woven fibers, mat, paper; can be composite or monolithic structure.

Thin material implies low cost.

Potential Areas of Concern

Pressure drop may be high.

Dead area in corners wastes heat exchange area.

Many seal surfaces - high probability of leakage.

MEMBRANE/MOLECULAR SIEVE   PROCESS

* HIGH TEMP. MEMBRANE/MOLECULAR SIEVE ENRICHES COMBUSTION AIR IN O_2
* USE EXCESS N_2 TO REACT WITH EXHAUST TO CLEAN
* ADSORBENT/ABSORBENT PARTICLES/PELLETS
   - ADSORB (ON SURFACE) OR ABSORB (WITHIN VOLUME) N_2 WHEN PREHEATING AIR: COOLING STATE
   - RELEASE N_2 WHEN HEATED IN THE EXHAUST STREAM: OFFGASSING MAY MINIMIZE FOULING ON PELLETS
   - USED IN CONJUNCTION WITH FLUIDIZED/RAINING BED TECHNOLOGY

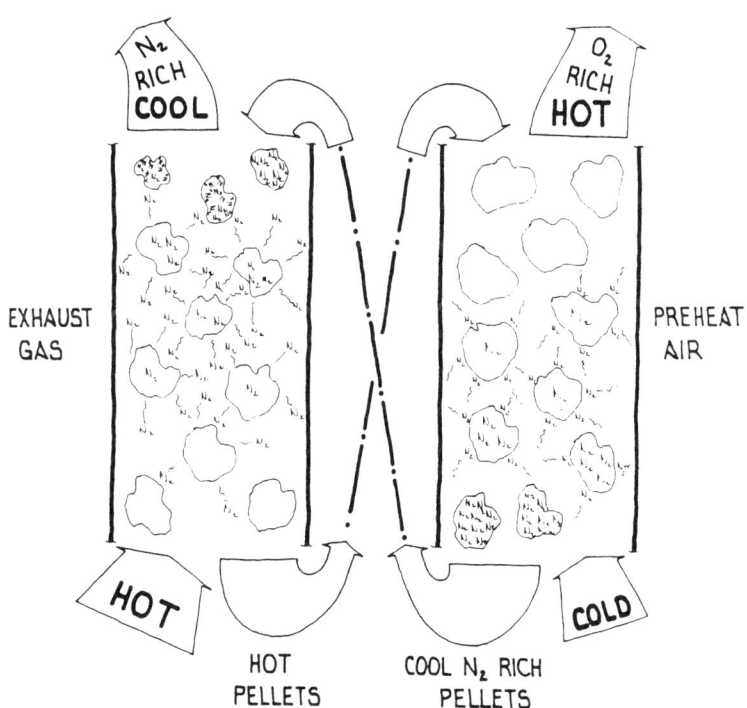

O_2 MOLECULAR SIEVE

Potential Advantages

Enriches combustion air in addition to exchanging heat.

Would be used with packed/raining fluidized bed.

Potential Areas of Concern

Material - Does not exist yet.

Critical temperature range for affinity change.

PREHEAT PELLETIZED FEEDSTOCK PROCESS

* SMALL PARTICLES OF FEEDSTOCK COULD BE USED AS MEDIA IN FLUIDIZED/RAINING BED Hx
* PREHEATED PELLETIZED FEEDSTOCK REDUCES RESIDUAL TIME IN PROCESS CONTAINMENT

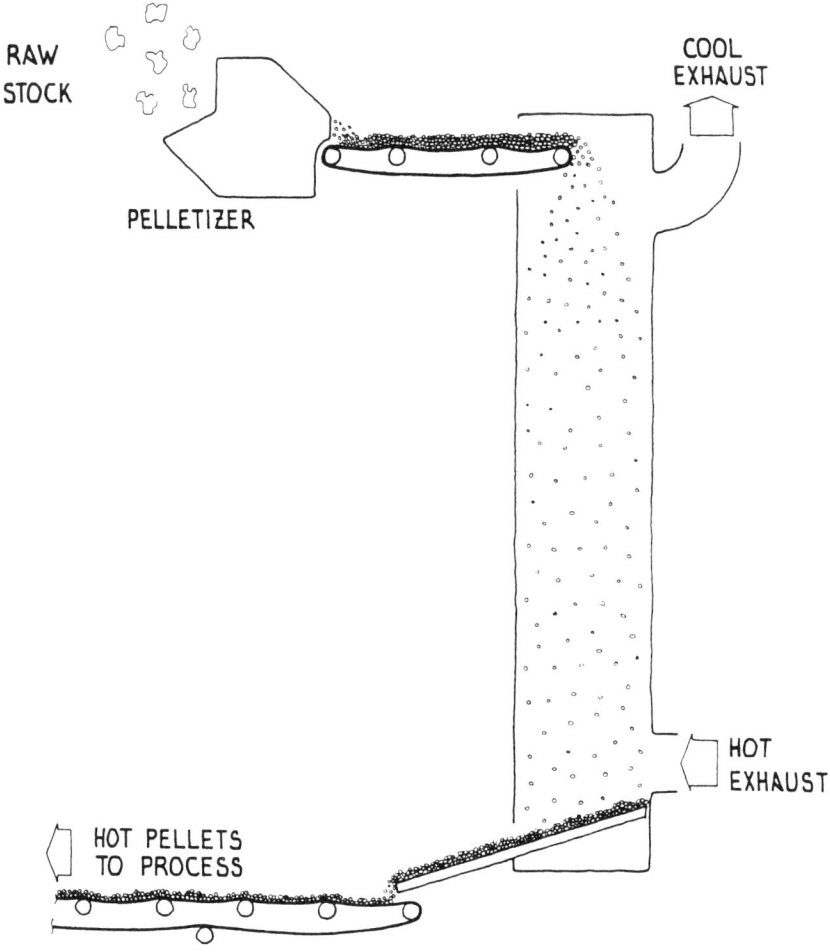

PREHEAT PELLETIZED FEEDSTOCK

Potential Advantages

Less residual time of stock in furnace - faster throughput.

Could be used with packed/raining fluidized bed.

Potential Areas of Concern

Does not provide preheated air for burners.

Only works for aluminum remelt and glass charging - does not work on steel reheat or glass tempering/annealing.

Needs high temperature distribution and control mechanisms.

: Thermal Electron is involved with this concept at a lower temperature range.

FLOATING SOLIDS CHANGE SHAPE PROCESS

* SOLID SHAPES HAVE HIGH SURFACE AREA TO VOLUME RATIO AND ARE MADE FROM MATERIALS WITH 'MEMORY'
* HIGH DRAG SOLIDS ARE SUSPENDED, ENTRAINED, OR FLOAT IN GAS STREAM — WHEN HOT THEY CHANGE SHAPE (LOW DRAG CONFIGURATION)
* LOW DRAG SOLID DROPS OUT OF GAS STREAM AND IS CONVEYED TO PREHEAT AIR STREAM
* COOL PARTICLE RETURNS TO HIGH DRAG SHAPE (MEMORY FEATURE)

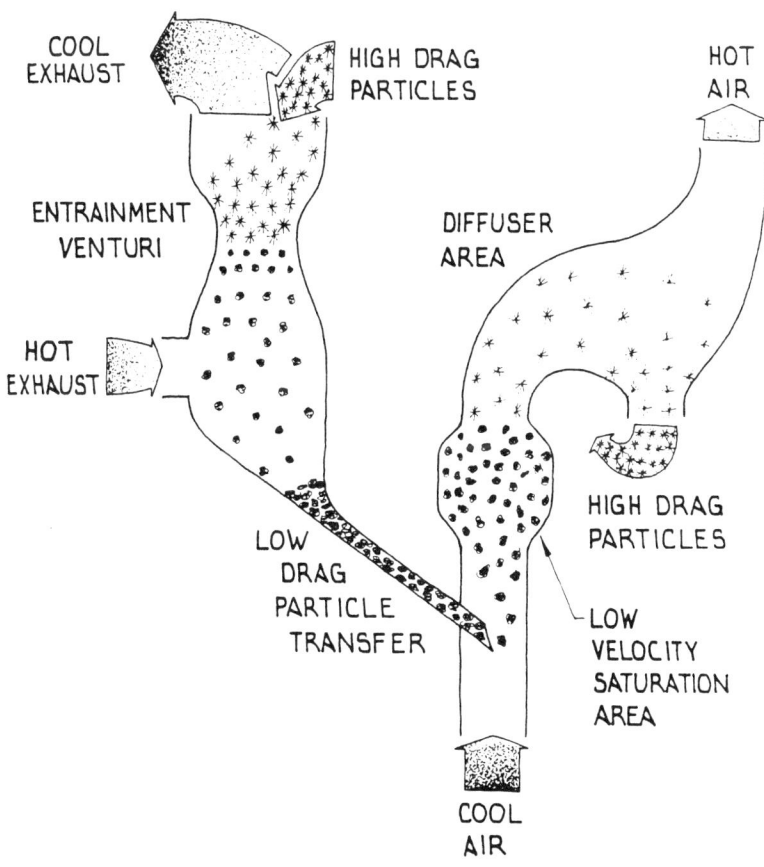

FLOATING SOLIDS CHANGE SHAPE

Potential Advantages

Lower pressure drop than packed bed.

Self-determined residual times for efficient heat transfer.

Potential Areas of Concern

Material not available yet.

Shape changing material may foul easily.

Not as compact as packed bed.

High temperature distribution and control system needed.

COGENERATION PROCESS

* USE THERMIONIC STACK WALLS TO GENERATE ELECTRICITY
* USE EXCESS PREHEAT AIR OR COOLED EXHAUST AS HEAT SOURCE FOR LOW TEMP. MHD, BRAYTON CYCLE TURBINE, OR DENSITY WHEEL

COEGENERATION: LOW TEMPERATURE MHD

Potential Advantages

Reclaims heat and produces other energy sources.

Potential Areas of Concern

Does not recuperate heat for reuse in process.

Reclaiming method for input into MHD technique unclear.

Materials uncertainties.

: This type of MHD (low-temp.) work is currently being done in Israel.

COGENERATION: THERMIONICS

Potential Advantages

Direct transfer of heat to electricity.

Fouling may not affect performance.

Potential Areas of Concern

Does not recuperate heat for reuse in process.

Materials uncertainties for thermionic walls.

Large size may be a problem.

Efficiency is unknown.

CHEMICAL REACTION PAIRS   PROCESS

* ENDOTHERMIC REACTION IN EXHAUST STREAM
* EXOTHERMIC TO RECOVER HEAT
* COULD BE CATALYST ACTIVATED EXOTHERMIC
* COULD BE USED IN A DISSOCIATION/RECOMBINATION METHOD
* COULD BE ENDOTHERMICALLY STABLE FOR SOME TIME PERIOD AFTER HEAT REMOVAL AND THEN REVERT TO ORIGINAL STATE BY UNDERGOING AN EXOTHERMIC REACTION

CHEMICAL REACTION PAIRS

Potential Advantages	Potential Areas of Concern
Compact size.	Materials (compounds) unavailable yet.
Possible filtering effect on waste stream.	High temperature collection, separation, and distribution system needed.
High heat transfer possible with finely dispersed spray.	Narrow temperature range for association and disassociation.
	Recombination area and method uncertain.

Appendix E

Phase Shot Tower Design Calculations

TABLE E.1. PHASE-CHANGE SHOT TOWER HX 400-FT TOWER

Cost Estimate Summary	
Develop Conceptual Design	$170,000
Design	500,000
Project Manager	150,000
Construction	3,445,000
Escalation	575,000
Contingency	<u>1,961,000</u>
TOTAL	$6,800,000

TABLE E.2. PHASE-CHANGE SHOT TOWER HX 150-FT TOWER

Cost Estimate Summary	
Develop Conceptual Design	$170,000
Design	500,000
Project Manager	90,000
Construction	2,240,000
Escalation	405,000
Contingency	<u>1,395,000</u>
TOTAL	$4,800,000

TABLE E.3. YEARLY OPERATING AND MAINTENANCE COSTS FOR THE 150-FOOT AND 400-FOOT PHASE-CHANGE SHOT TOWER HEAT EXCHANGERS

Maintenance[a]	150-Foot	400-Foot
Conveyors	$ 95,000	$230,000
Vib. Screen	17,000	17,000
Heaters	81,000	81,000
Man Lift	41,000	61,000
Sieve	110,000	110,000
Jib Crane	129,000	129,000
CAPITAL COST	$473,000	$628,000
Yearly Maintenance	$ 94,600	$125,000
Operating Cost[b]		
Labor	$156,000	$156,000
Electricity		
HP for 3 Conveyors	90	150
HP for Vib. Screen	10	15
	100 HP	165 HP
Cost for Power	$ 29,000	$ 48,000
TOTAL YEARLY OPERATING AND MAINTENANCE COSTS	$310,000	$380,000

a. Calculated as 20% of the capital cost of the equipment in severe service.

b. Assuming 325 operating days per year, 1 man per shift, $20 per hour, and 0.05¢ per kilowatt-hour.

ENGINEERING DESIGN FILE

EG&G Idaho, Inc.
FORM EG&G-2631 (Rev. 4-78)

PROJECT FILE NO. _____
EDF SERIAL NO. _____
FUNCTIONAL FILE NO. 613004400

PROJECT/TASK: AHET
DATE: _____
SUBTASK: Phase-change shot tower heat exchanger
EDF PAGE NO. 1 OF ___

SUBJECT

ABSTRACT

DISTRIBUTION (COMPLETE PACKAGE):

DISTRIBUTION (COVER SHEET ONLY): PROJECT EDF FILE LOG, EDF SERIAL NO. LOG

AUTHOR	DEPT.	REVIEWED	DATE	APPROVED	DATE

Appendix E: Phase Shot Tower Design Calculations

AHET EDF p.2

	HIGH FIRE	LOW FIRE
AIR SIDE FLOW	144.3 LBM/MIN	48.1 LBM/MIN
GAS SIDE FLOW	151.9 LBM/MIN	50.6 LBM/MIN
AIR INLET TEMP.	100 °F	100 °F
GAS INLET TEMP.	2450 °F	2450 °F
AIR OUTLET TEMP.		2000 °F
TURNDOWN RATIO	3 TO	1
AIR SIDE ΔP	30 IN of H₂O	
GAS SIDE ΔP	8 IN of H₂O	

From ②
Desirable characteristics of the Recuperator

 Idaho, Inc.

FORM EG&G-1592
(Rev 5-77)

CALCULATION WORK SHEET

Subject _Phase-change shot tower: Schematic_ Page __3__ of ____ Pages
Date _6-23-83_
Prepared By _____ Checked _____ Work Request _6 2 504 400_

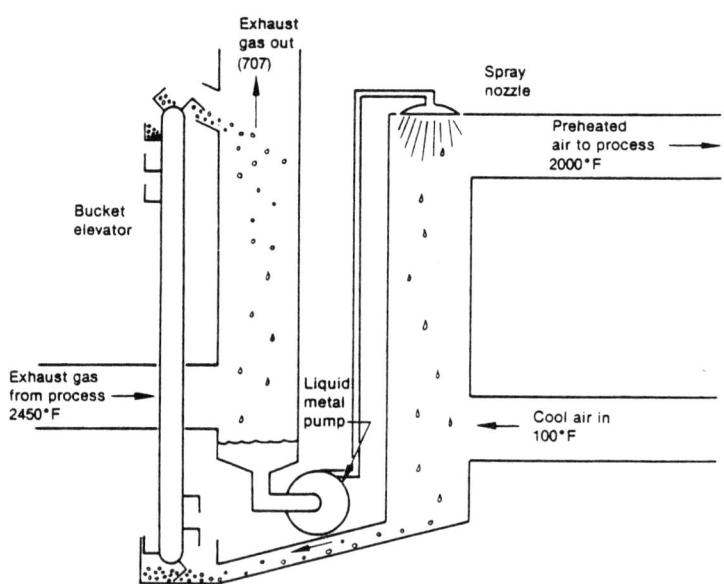

Appendix E: Phase Shot Tower Design Calculations 177

 EG&G Idaho, Inc.

FORM EG&G-1592
Rev 5-77

CALCULATION WORK SHEET

Page __4__ of _____ Pages

Subject __Phase-change shot tower HX: Alternate Schematic__ Date __6-28-83__

Prepared By _____ Checked _____ Work Request __G13 004 400__

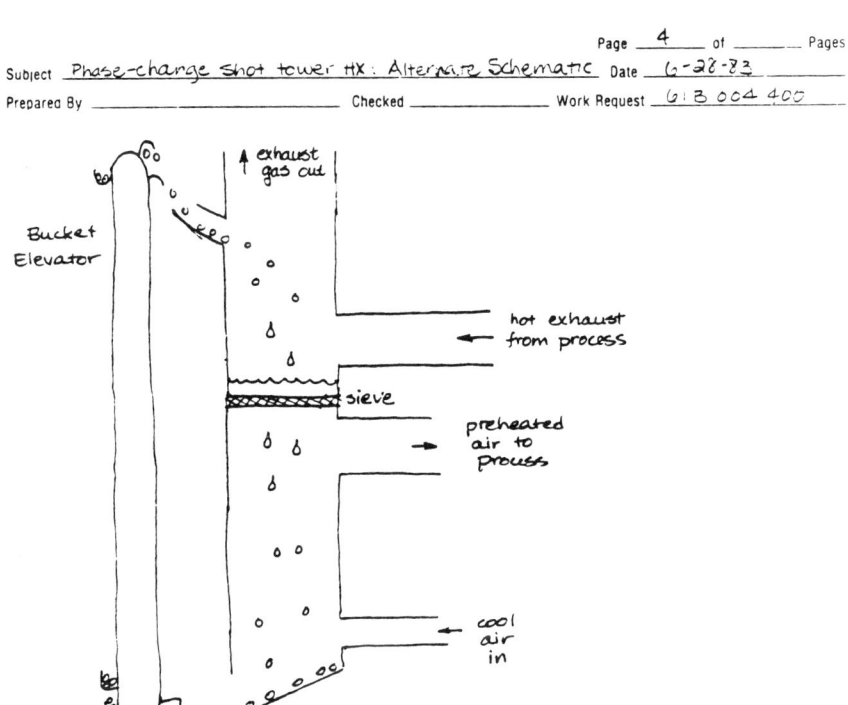

5-23-?3
p 5

Low fire conditions:

$\dot{m}_g = 50.6$ lbm/min

$\dot{m}_a = 48.1$ lbm/min $\Delta T = 2000 - 100 = 1900°F$

use C_p gas $= .286$ Btu/lbm°F

C_p air $= .276$ Btu/lbm°F

$(\dot{m} C_p \Delta T)_{gas} = (\dot{m} C_p \Delta T)_{air} = \dot{Q}$

$(50.6)(.286)(2450 - T_{ex}) = (48.1)(.276)(1900) = \dot{Q}$ Btu/min

$T_{ex} = 707°F$

$\dot{Q} = 25,200$ Btu/min

$\dot{Q}$ is the rate at which heat must be transferred to or from the metal.

$$\dot{Q} = \dot{m}_m \left[C_{p\,LIQ}(T_H - T_m) + h_{fg} + C_{p\,SOL}(T_m - T_c) \right]$$

where

$\dot{m}_m$ = mass flowrate of metal

$C_{p\,LIQ}$ = specific heat of the liquid metal

T_H = temperature of the liquid metal

T_m = melting temperature

h_{fg} = latent heat of melting

$C_{p\,SOL}$ = specific heat of the solid metal

T_c = temperature of the solid metal brass

T_c can't be less than $100°F$

T_H can't be more than $2450°F$

Appendix E: Phase Shot Tower Design Calculations

6-27-83

p. 6

High fire conditions:
$\dot{m}_g = 151.9$ lbm/min
$\dot{m}_a = 144.3$ lbm/min

$(151.9)(.286)(2450 - T_{ex}) = (44.3)(.276)(1900) = \dot{Q}$
$\dot{Q} = 75700$ Btu/min
$T_{ex} = 702\,°F$

 EG&G Idaho, Inc.

FORM EG&G-1592
(Rev 5-77)

CALCULATION WORK SHEET

Subject: Size of raining-bed Hx
Page 7 of ___ Pages
Date 6-27-83
Prepared By ___ Checked ___ Work Request L13 004 455

References: R. Decher, "Falling-Bead Dry Cooling Tower"
Perry, Chemical Engineers' Handbook

We need to size a raining-bed heat exchanger for the melting half of the phase-change shot tower heat exchanger. (For now, ignore phase change.)

Need: Height of tower
Diameter of tower
Working fluid flowrate (assume aluminum)

From Decher's paper, the tower height is

$$H = \frac{\delta}{K} \frac{\Delta T_G}{\Delta T} \left(\frac{1}{1 + u_o/V_o} \right) r_o$$

The tower cross-sectional area is

$$A_o = Q \left(C_p \Delta T_G \left[1 + \frac{u_o}{V_o} \right] \rho \cdot u_o \right)^{-1}$$

Appendix E: Phase Shot Tower Design Calculations   181

EG&G Idaho, Inc.
FORM EG&G-1592
Rev 5-77

CALCULATION WORK SHEET

Subject: Reference bead radius (r_o^*)
Page 8 of ___ Pages
Date 6-27-83
Prepared By ___ Checked ___ Work Request 613 054 450

Choose bead radius such that $\tau_1 = \tau_2$ at $\Delta T_G/\Delta T = 1$
  τ_1 = time scale for conduction of heat from bead surface to bead center
  τ_2 = time scale for beads to fall the height of the tower

$$\tau_1 \sim \frac{r_o^2}{\alpha}$$

$$\tau_2 = \frac{\Delta T_G}{\Delta T} \sqrt{\delta \frac{r_o}{g}} \frac{1}{k}$$

$g = 32.2 \text{ ft/s}^2 = 386.4 \text{ in/s}^2$
$\delta = 11400$
$k = 1.20$

set $\tau_1 = \tau_2$ $\Delta T_G/\Delta T$ is set equal to 1 to find r_o^*

$$\frac{r_o^2}{\alpha} = \frac{\Delta T_G}{\Delta T} \sqrt{\delta \frac{r_o}{g}} \frac{1}{k}$$

$$\alpha = \frac{k}{\rho c} = \frac{\left(1540 \frac{Btu\cdot in}{ft^2\cdot hr\cdot °F}\right)}{\left(0.098 \frac{lb}{in^3}\right)\left(0.23 \frac{Btu}{lb\cdot °F}\right)} \left(\frac{1}{144} \frac{ft^2}{in^2}\right) = 474 \frac{in^2}{hr}$$

$$r_o^{1.5} = \frac{\alpha}{k} \frac{\Delta T_G}{\Delta T} \sqrt{\frac{\delta}{g}}$$

$$r_o^* = \left[\frac{474 \text{ in}^2/hr}{1.20} \;(1)\; \left(\frac{11400}{386 \text{ in/s}^2}\right)^{1/2} \; \frac{1}{3600} \frac{hr}{s} \right]^{.67}$$

Reference bead radius $r_o^* = 0.71$ in.

This was how F. Decker calculated the bead radius for glass beads. Larger beads (radius > r_o^*) will not heat through fast enough. Smaller beads will work.

Idaho, Inc.

FORM EG&G-1592
Rev 5-77

CALCULATION WORK SHEET

Page __9__ of ____ Pages

Subject __Tower height__   Date __6-27-83__

Prepared By _____ Checked _____ Work Request __613 004400__

Evaluate tower height

δ = drag/density parameter = 11,400 for aluminum

$K = C_{p,air} / C_{p,bead}$

$C_{p,air}$ = 0.276 Btu/lbm °F

$C_{p,aluminum}$ = 0.23 Btu/lbm °F   } $K = 1.20$

ΔT_a = temperature change experienced by the beads or the air, 1743°F (exhaust gas temp change = 2450 - 707 °F)

ΔT = temperature difference (constant) between beads and air, assumed to be 30°F

u_o = air velocity (evaluated at air inlet)

V_o = bead velocity (constant)

for aluminum spheres, the terminal velocity = $V_t = 605 \sqrt{r_o}$ $\frac{+}{5}$ relative to the air; r_o is the sphere radius in feet.

∴ $V_o = 605 \sqrt{r_o} - u_o$

Choose air velocity $u_o = 30$ ft/s and $r_o = 0.71$ in = 0.059 ft

∴ $V_o = 117$ ft/s

$H = \dfrac{11400}{1.20} \dfrac{1743}{30} \left(\dfrac{1}{1 + \frac{30}{117}} \right) 0.059$ ft = 25900 ft ≈ 5 miles (too high)

Use a smaller bead radius: $\frac{1}{16}$" = 0.00521 ft

For $u_o = 30$ fps, $V_o = 43.7 - 30 = 13.7$ fps

$H = \dfrac{11400}{1.2} \dfrac{1743}{30} \left(\dfrac{1}{1 + \frac{30}{13.7}} \right) 0.00521 = 902$ ft   still too high

Use a smaller approach temperature: change ΔT from 30 to 150

$H = 902 \left(\dfrac{30}{150} \right) = \underline{\underline{180 \text{ feet}}}$

Appendix E: Phase Shot Tower Design Calculations

FORM EG&G-1592
Rev 5-77

CALCULATION WORK SHEET

Subject: Tower area (first cut)
Page 10 of ___ Pages
Date: 6-28-73
Prepared By: ___ Checked: ___ Work Request: 6-3-064-472

Evaluate tower area

$$\frac{A_c}{Q} = \left(C_p \Delta T_G \left[1 + \frac{u_c}{V_o} \right] \rho_o u_o \right)^{-1}$$

$$\frac{A_c}{Q} = \left(0.276 \cdot 1743 \left[1 + \frac{30}{13.7} \right] .0208 \cdot 30 \right)^{-1}$$

$$\frac{A_c}{Q} = .00104 \; \frac{ft^2}{Btu/s}$$

$C_p = 0.276 \; Btu/lbm \cdot °F$
$\Delta T_G = 1743 \; °F$
$u_c = 30 \; ft/s$
$V_o = 13.7 \; ft/s$
$\rho_o = .0208 \; lb/ft^3$

Low fire conditions: $Q = 25200 \; \frac{Btu}{min} = 420 \; \frac{Btu}{s}$

∴ area $A = .44 \; ft^2$ diameter $= .75 \; ft = 9"$

Does this agree with the gas flow we need?

gas flow (low fire) $\dot{m} = 50.6 \; lbm/min$

$\rho = .0208 \; lbm/ft^3$

if $u_o = 30 \; ft/s$:

$$A = \frac{\dot{m}}{\rho \, u_o} = \frac{(50.6 \; \frac{lbm}{min})(\frac{1}{60} \; \frac{min}{sec})}{(.0208 \; \frac{lbm}{ft^3})(30 \; \frac{ft}{sec})} = 1.35 \; ft^2$$

∴ The assumption of 30 ft/s gas velocity should be changed

EG&G Idaho, Inc.

FORM EG&G-1592
(Rev 5-77)

CALCULATION WORK SHEET

Subject __1/16" radius aluminum bead__   Page __11__ of ___ Pages
Prepared By _____ Checked _____   Date __6-23-83__
                                            Work Request __613 604 400__

$$A_o = \frac{\dot{m}}{\rho u_o} = Q \left(c_p \Delta T_G \left[1 + \frac{u_o}{V_o} \right] \rho, u_o \right)^{-1}$$

$$\dot{m} = Q \left(c_p \Delta T_G \left[1 + \frac{u_o}{V_o} \right] \right)^{-1}$$

$$\left(50.6 \; \frac{lbm}{min} \right) = (25200 \; \frac{Btu}{min}) \left[\left(.276 \; \frac{Btu}{lbm\,°F} \right) (1743°F) \left[1 + \frac{u_o}{V_o} \right] \right]^{-1}$$

$$\left[1 + \frac{u_o}{V_o} \right] = \frac{(25200)}{(50.6)(.276)(1743)} = 1.035$$

$$\frac{u_o}{V_o} = .035$$

for a 1/16" radius particle, $V_o = 43.7$ fps $- u_o$

$$\frac{u_o}{43.7 - u_o} = .035 \qquad u_o = \frac{(.035)(43.7)}{1.035}$$

$u_o = 1.49$ ft/s $V_o = 42.2$ ft/s

$A = 27.2$ ft²   diameter = 5.9 feet = 5'10½"

This also changes the tower height:

$$H = \frac{11400}{1.2} \cdot \frac{1743}{150} \left(\frac{1}{1 + \frac{1.49}{42.2}} \right) 0.00521 = 556 \text{ feet}$$

To get the tower height down, we could
1. Reduce the ΔT_G (i.e. gas leaves at $T > 707$)
2. Increase approach temperature (ΔT btw bead and air)
3. Decrease bead size

Would like to decrease H by a factor of 2.78 (to $H \approx 200$ feet)

Appendix E: Phase Shot Tower Design Calculations

EG&G Idaho, Inc.

FORM EG&G-1592
Rev 5-77

CALCULATION WORK SHEET

Page __12__ of ____ Pages
Subject __Aluminum bead; radius = 1/64"__   Date __6-23-83__
Prepared By _____ Checked _____ Work Request __G1B004400__

To cut the PU bead size down. This will also affect u_0 and V_0.

Bead radius = $\frac{1}{64}"$ = .01563" = .00130 ft

$V_0 = 605\sqrt{T_0} - u_0 = 21.83 - u_0$

$$\frac{u_0}{V_0} = \frac{Q}{m\, C_p\, \Delta T_0} - 1$$

Low fire conditions:

$$\frac{u_0}{V_0} = \frac{25200}{(50.6)(276)(1743)} - 1 = .035 = \frac{u_0}{21.8 - u_0}$$

$$u_0 = \frac{(.035)(21.8)}{1.035} = 0.738 \text{ ft/s}$$

$V_0 = 21.09$ ft/s

$$H = \frac{11400}{1.2} \cdot \frac{1743}{150} \left(\frac{1}{1.035}\right) \cdot .00130 = 139 \text{ feet} \quad \text{tower height}$$

$$A_0 = \frac{m}{\rho u_0} = \frac{(50.6)(\frac{1}{60})}{(.0208)(0.738)} = 54.9 \text{ ft}^2 \quad (\text{diameter} = 8.36 \text{ feet})$$

EG&G Idaho, Inc.

FORM EG&G-1592
(Rev 5-77)

CALCULATION WORK SHEET

Subject: High fire conditions (Aluminum; $r_b = \frac{1}{24}''$)   Page 13 of ___ Pages
Date 6-28-83
Prepared By ___   Checked ___   Work Request GIB 004 403

For high fire conditions, the gas side flow approximately triples ($\dot{m} = 151.9$ lbm/min) as does the heat transfer rate ($Q = 75500$). Sizing the tower based on this gas flow:

$$\left[1 + \frac{u_o}{V_o}\right] = 1.035 \cdot \frac{50.6}{151.9} \cdot \frac{75700}{25200} = 1.035$$

∴ the height of the tower stays the same (139 feet)

$$A = \frac{\dot{m}}{\rho u_o} = 3(54.9 \text{ ft}^2) = \underline{165 \text{ ft}^2}$$ (Area must triple to accommodate the increased gas flow)

What will happen at low fire if we size the tower area for high fire?
- Air flow velocity u_o will decrease by a factor of 3
- Bead velocity V_o will increase (∴ fall faster, lose less heat)
- The required tower height will increase

$u_o = \frac{1}{3} \cdot 0.738 = 0.246$

$V_o = 605 \sqrt{r_b} - 0.246 = 21.6 \text{ ft/s}$

$\frac{u_o}{V_o} = 0.0114$

$1 + \frac{u_o}{V_o} = 1.0114$

$H = 139 \cdot \frac{1.035}{1.0114} = \underline{142 \text{ feet}}$

Assume we can size the tower __area__ for high fire conditions (high gas flowrate) and size the tower __height__ for low fire conditions.

CALCULATION WORK SHEET

Subject: Advantages & disadvantages
Page 14 of ___ Pages
Date 6-28-83
Prepared By ___ Checked ___ Work Request 618 004 600

Advantages of this design over a raining-bed HX:
- The beads are cool when they are in the bucket elevator ∴ the material of the bucket elevator can be less expensive
- The beads can store more heat because the latent heat of melting/freezing is used (Δh_{fg}) as well as the sensible heat ($\Delta h = C_p \Delta T$)

Disadvantages
- A liquid metal pump will probably be needed
- Aluminum will oxidize, any working fluid will pick up contaminants from exhaust gases
- Cooling for the liquid metal pump is required
- Not sure if liquid metal pump can be made of refractory
- Pump would be inefficient
- If a gravity system is used, the tower would be very tall ∴ expensive and maybe violating zoning laws
- Sieve would get blocked by exhaust particulates
- We know of no way to pump glass
- We don't know if a sieve would work anyway
    - would the droplets be the right size
    - would the material flow through the sieve fast enough

 EG&G Idaho, Inc.

CALCULATION WORK SHEET

Subject: Working Fluid Flowrate (Aluminum)   Page 15   Date 6-23-83   Work Request 618 CC4 400

Aluminum flowrate:

$$Q = \dot{m}\left[C_{P\ell}(T_H - T_m) + \Delta h_{fg} + C_{PS}(T_m - T_c)\right]$$

$Q = 65200$ Btu/min low fire ; 75700 Btu/min high fire
$\dot{m}$ = aluminum mass flowrate
$C_{P\ell}$ = specific heat of molten aluminum =
C_{PS} = specific heat of solid aluminum =
Δh_{fg} = latent heat of melting = 94.5 cal/g = 170 Btu/lbm
T_H = highest temperature of Al =
T_c = lowest temperature of Al =
T_m = melting point of Al = $1220.4°F$

Aluminum (or any metal) will oxidize when it drops through the oxygen-bearing preheat air. An oxide layer on the outside of the metal beads will inhibit heat transfer on the melting side.

∴ Look at using glass as the working fluid.

Appendix E: Phase Shot Tower Design Calculations

 Idaho, Inc.

FORM EG&G-1592
(Rev 5-77)

CALCULATION WORK SHEET

Page ___16___ of _____ Pages
Subject Glass $\tfrac{1}{32}$" radius
Date 6-22-83
Prepared By _____ Checked _____ Work Request 613 664 400

For $\tfrac{1}{32}$" bead radius, evaluate tower area at high fire conditions and tower height at low fire conditions.

$\delta = 11400$
$K = C_{p,air} / C_{p,glass} = 0.276 / 0.16 = 1.73$ ($C_p = .12$ to $.20$ for glass)
$\Delta T_a = 1743$ Perry Eng. manual
$\Delta T = 150°$ (assumed approach temperature)
$r_o = \tfrac{1}{32}$" assumed $= .0026$ ft
$\therefore V_o = 605\sqrt{r_o} - u_o = 30.87 - u_o$

$\dfrac{u_o}{V_o} = \dfrac{Q}{\dot{m} C_p \Delta T_a} - 1 = 1.035$ at low fire or high fire

$u_o = \dfrac{(.035)(30.87)}{1.035} = 1.044$ ft/s

$V_o = 30.87 - 1.044 = 29.8$ ft/s

Tower area at high fire $= \dfrac{\dot{m}}{\rho u_o} = \dfrac{151.9 \tfrac{lbm}{min} \cdot \tfrac{1}{60} \tfrac{min}{sec}}{.0208 \tfrac{lbm}{ft^3} \cdot 1.044 \tfrac{ft}{s}} = 117$ ft²

$\underline{12.2'\ diameter}$

$\therefore$ at low fire, $U_o = \tfrac{1}{3}(1.044) = 0.348$ and $V_o = 30.53$

$H = \dfrac{11400}{1.73} \cdot \dfrac{1743}{150} \left(\dfrac{1}{1.0114}\right)(.0026) = \underline{197\ feet}$

Use a smaller bead to get a smaller tower height

EG&G Idaho, Inc.

FORM EG&G-1592
(Rev 5-77)

CALCULATION WORK SHEET

Subject __Glass $\frac{1}{2}$" radius__   Page __17__ of ____ Pages
Prepared By ____ Checked ____ Date __6-29-83__ Work Request ____

For $\frac{1}{64}$" bead radius $r_o = \frac{1}{64}" = .0156" = .00130$ ft

$V_s = 21.83 - u_o$

$u_o = \frac{(.035)(21.83)}{1.035} = .738$ ft/s

$V_o = 21.09$ ft/s

area (high fire) $A = 117 \cdot \frac{1.044}{.738} = 165$ ft^2 (d = 14.5')

at low fire, $u_o = \frac{1}{3}(.732) = .246$ ft/s

$V_o = 21.58$ ft/s $\left.\begin{array}{c} \\ \\ \end{array}\right] \frac{u_o}{V_o} = .0114$

$H = \frac{11400}{1.73} \cdot \frac{1743}{150} \left(\frac{1}{1.0114}\right) \cdot .00130 = 98$ feet

The approach temperature ΔT was assumed to be 150°. Actually, the sieve will have very little loss because it is thin and completely enclosed in the heat exchanger.
Assume $\Delta T \approx (2450 - 2000)/2 = 225°F \rightarrow$ use 200 °F

∴ Tower height is reduced $H = 98 \cdot \frac{150}{200} = \underline{\underline{74 \text{ ft}}}$

Appendix E: Phase Shot Tower Design Calculations

EG&G Idaho, Inc.

CALCULATION WORK SHEET

Page 18 of _____ Pages
Subject: CFM
Date: 6-29-83
Prepared By: _____ Checked: _____ Work Request: 613 004 40

Cost estimating requested ACFM of gas & air streams.

AIR

Low fire: air flow = 43.1 lbm/min $T_{in} = 100°F$ $T_{out} = 2000°F$
$P_{in} = .0709 \frac{lbm}{ft^3}$ $P_{out} = .01213 \frac{lbm}{ft^3}$

$ACFM_{in} = 678 \approx 700$
$ACFM_{out} = 3965 \approx 4000$

High fire: air flow = 144.3 lbm/min $ACFM_{in} = 2035 \approx 2000$
$ACFM_{out} = 11896 \approx 12000$

GAS

Low fire: gas flow = 50.6 lbm/min $T_{in} = 2450°F$ $T_{out} = 707°F$
$P_{in} = .00486$ $P_{out} = .03454$

$ACFM_{in} = 10412 \approx 10000$
$ACFM_{out} = 1465 \approx 1500$

High fire: gas flow = 151.9 lbm/min $ACFM_{in} = 31255 \approx 31000$
$ACFM_{out} = 4398 \approx 4400$

EG&G Idaho, Inc.

FORM EG&G-1592
(Rev 5-77)

CALCULATION WORK SHEET

Page __19__ of ____ Pages
Subject __Working Fluid (Glass) Flowrate__   Date __6-29-83__
Prepared By _____ Checked _____ Work Request __GIB CO4 400__

The heat capacity of glass generally increases with temperature (see diagram on next page for SiO_2). There is no well-defined melting point. Assume there's no latent heat of fusion (worst case: if there IS a latent heat of fusion, the tower would be shorter). Glass temperature will change between 2250°F and 300°F approximately.

Total heat to be transferred is 25200 Btu/min Low Fire
                                75700 Btu/min High Fire

Break the heat transfer into two increments of ΔT. Integrate (dT times the C_p of the glass) in each interval.

The transition temp. (on Fig 12.3) occurs at ~ 870°K = 1110°F

Interval	T (°F)	Eqn for C_p		Points (T, C_p)
1	300 - 1100	$C_p = 8.83 \times 10^{-5} T + 0.194$		(261°F, .217) and (1110, .292)
2	1100 - 2250	$C_p = 2.73 \times 10^{-5} T + 0.236$		(1110, .266) and (2201, .292)

$(C_p$ is in $\frac{Btu}{lb\cdot°F})$

$\Delta H = C_p \Delta T$

$\Delta H_{total} = \int_{300}^{1110} C_p dT + \int_{1110}^{2250} C_p dT$

$= \left[8.83 \times 10^{-5} \frac{T^2}{2} + 0.194 T \right]_{300}^{1110} + \left[2.73 \times 10^{-5} \frac{T^2}{2} + 0.236 T \right]_{1110}^{2250}$

$= (54.421 - 3.975) + 157.094 + (69.203 - 16.843) + 268.644$

$= 528.544$ Btu/lb

So 1 lb of glass absorbs ~530 Btu to change temperature from 300°F to 2250°F.

High fire: $\dot{m} = 75700 / 530 = 143$ lb/min
Low fire: $\dot{m} = 25200 / 530 = 47.5$ lb/min

Appendix E: Phase Shot Tower Design Calculations

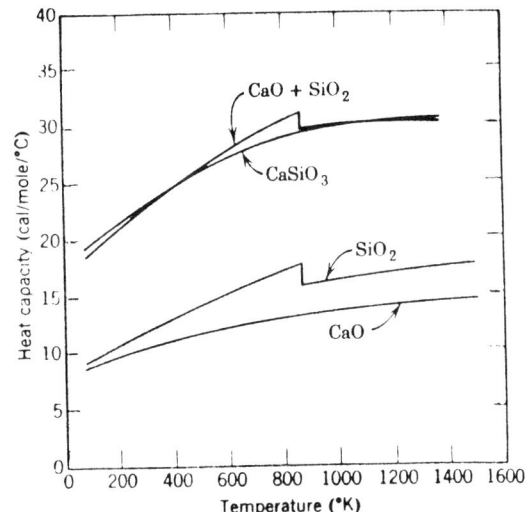

Heat capacity of various forms of Ca) + SiO_2 in 1:1 molar ratio.

CALCULATION WORK SHEET

Subject: Total charge of working fluid
Page: 21 of ___ Pages
Date: 6-29-83
Prepared By: ___ **Checked** ___ **Work Request:** 61B 004 400

For a total tower height of $2H = 2.197 = 394$ ft
The bead velocity $V_o = 30.5$ ft/s (at low fire)
∴ a bead takes 12.9 s to fall

Flowrate = 47.5 lb/min = .792 lb/s
∴ tower holds 10.2 lbs in free-fall   low fire
At high fire $V_o = 29.8$ ft/s and flowrate = 143 lb/min = 2.38 lb/s
∴ holdup is 31.5 lbs
___

For a total tower height of $2H = 2.74 = 148$ ft
Low fire: $V_o = 21.58$ ft/s   ∴ $t = 6.86$ s
Holdup = 5.4 lbs

High fire: $V_o = 21.09$ $t = 7.02$ s   Holdup = 16.7 lbs
___

Assume 3" of liquid on top of the sieve. $\rho = 162$ lb/ft^3

Tall tower ($2H = 394$ ft) $d = 12.2'$ $V = 9.3$ ft^3 $M = 1510$ lbs

Short tower ($2H = 148$ ft) $d = 14.5'$ $V = 13.1$ ft^3 $M = 2130$ lbs
___

Total charge of working fluid = tower holdup + holdup on sieve + reservoir of beads at the bottom + reservoir of beads in the hopper at the top + holdup in the bucket elevator

For the hoppers and conveyors, use the density of "broken glass, loose" = 2000 lb/cu yd = 74.1 lb/ft^3

From Handbook of Solid Waste Management, D.G. Wilson, 1977, page 42

Appendix E: Phase Shot Tower Design Calculations

FORM EG&G-1592
Rev 5-77

CALCULATION WORK SHEET

Page 22 of _____ Pages
Subject: Glass Inventory
Date: 3-1-83
Prepared By: _____ Checked: _____ Work Request: 61B 604 400

For the tall tower,
    ... screen size = 6×12. Assume 1 foot depth.
    Volume = 72 ft³
    Mass = 72 × 74.1 = 5335 lbs
    Hopper size = 16'T × 15' dia
    Volume = 2827 ft³
    Mass = 2827 × 74.1 = 209,500 lbs
    Conveyor (moving 143 lb/min @ high fire): use 9" conveyor so that the speed can be kept low, since glass is abrasive
    Volume = 400' tall × 9¾" × 6" × $\frac{1}{144}$ = 162.5 ft³
    Mass = 162.5 × 74.1 = 12,040 lbs

Total mass (tall tower) = 31.5 + 1510 + 5335 + 209,500 + 12,040 = **228,000 lbs**

For the short tower,
    Screen 5335 × $\frac{10}{12}$ = 4446
    Hopper (same) = 209,500 lbs
    Conveyor 12,040 × $\frac{150}{400}$ =

Total mass (short tower) = 16.7 + 2130 + 4446 + 209,500 + 4515 = **221,000 lbs**

196 Ceramic Heat Exchanger Concepts and Materials Technology

page 23

Capacities of SERIES 1000 REDLER® Conveyor/Elevators

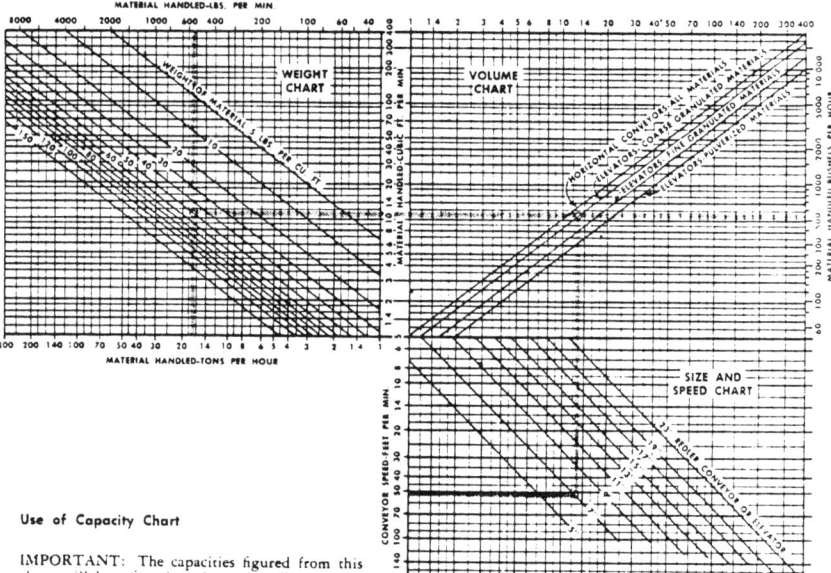

Use of Capacity Chart

IMPORTANT: The capacities figured from this chart will be approximately correct for all ordinary materials and conditions. Contact Stephens-Adamson in regard to exact capacities, materials that are sticky, very fragile or unusually abrasive and in regard to special applications such as high temperature operation.

To determine the capacity of a REDLER unit in tons per hour or pounds per minute, the weight of material will be a factor and we use the left hand Weight Chart, entering at either top or bottom and moving to the diagonal line for the weight of materials. We then move across to the right hand Volume Chart and convert the capacity into volume.

If the capacity is required by volume (cubic feet per minute or bushels per hour), weight is disregarded and we enter directly into the right hand section of the chart at the desired volume. Here we move horizontally to the diagonal line indicating the type of Conveyor or Elevator necessary to handle given material.

From this intersection we drop vertically into the lower Size and Speed Chart, where we intersect one or more diagonals that show the sizes of Conveyors and Elevators that will handle the volume required. To the left of each intersection we find the necessary speed for each size of Conveyor. For Loop Boot Elevators loaded from inside of loop, add 10% to Elevator speed.

EXAMPLE:

The shaded line shows the method of finding the size and speed of a REDLER Elevator to handle 17 tons of coal per hour. Enter the WEIGHT CHART at 17 tons per hour (or the equivalent 575 lbs. per min.) and move up to the diagonal indicating coal a 50 lbs. per cu. foot. The line then moves horizontally into the right hand VOLUME CHART, to show a volume of 12 cu. feet per min. From the intersection of this line with the diagonal "Elevators Handling Granulated Materials." the line drops to the diagonal "7" Conveyor or Elevator" in the SIZE CHART. Then left to the recommended speed of 50 F.P.M. The same procedure also shows a 9" Elevator at 35 F.P.M. or a 5" Elevator at 90 F.P.M.

Conveyor size	5	7	9	11	13	15	17	19	23
duct dimensions - (inches)	5½ / 4	7¾ / 5¼	9¾ / 6	12 / 7⅞	14 / 8	16 / 9	18 / 10½	20 / 12	24 / 15

Appendix E: Phase Shot Tower Design Calculations

CALCULATION WORK SHEET

Page __24__ of ____ Pages
Subject __Problems__ Date __6-29-83__
Prepared By _____ Checked _____ Work Request __613 004 400__

more problems:
- molten glass will adhere & solidify on the tower walls (maybe we can install baffles to swirl air & keep the glass off the walls)
- Pollutants may be acid & erode the firebrick

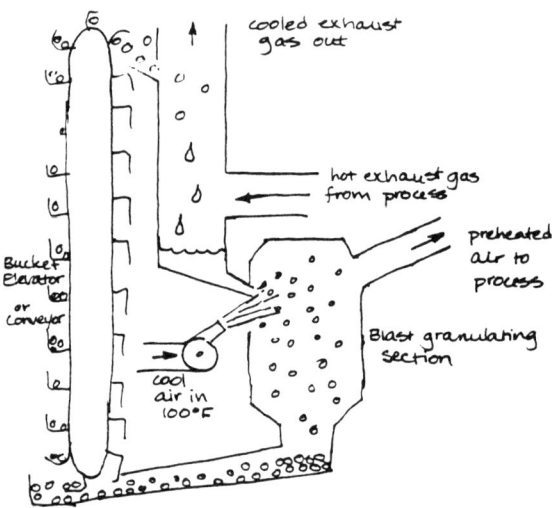

Melting-bead / Blast granulating heat exchanger

this system alone can't preheat the air to a high temperature or cool the beads to a low temperature: at best the beads and air will be the same temperature.

Air flow = 144.3 lbm/min C_p = .276 Btu/lbm°F T_{in} = 100°F
WF flow = 143 lbm/min C_p = .292 Btu/lbm°F T_{in} = 2200°F

$\dot{m} C_p \Delta T_{air} = \dot{m} C_p \Delta T_{WF}$

$(144.3)(.276)(T_{out} - 100) = (143)(.292)(2200 - T_{out})$

T_{out} = 1175 °F

Appendix E: Phase Shot Tower Design Calculations 199

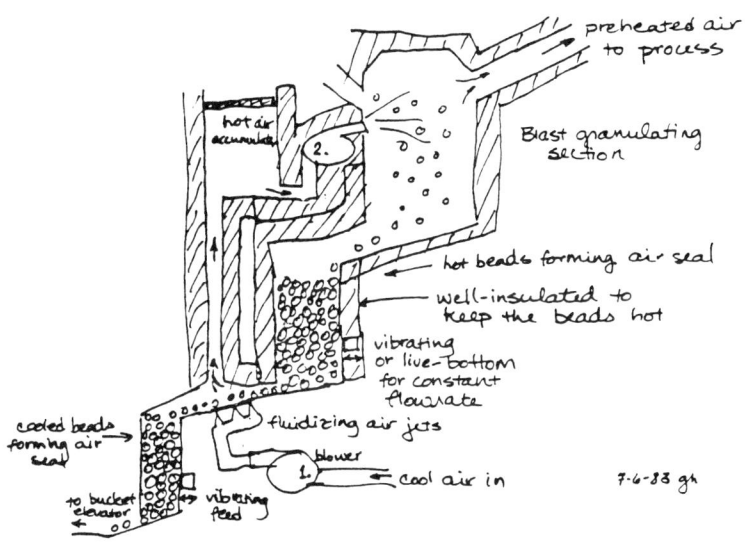

Assume that the air can be heated to 600°F and still be handled by the blower (#2).

In the blast granulating section,

$$(144.3 \tfrac{lbm}{min})(.276 \tfrac{Btu}{lbm\,F})(T_{out} - 600) = (143 \tfrac{lbm}{min})(.292)(2200 - T_{out})$$

$T_{out} = 1420°F$ This is the temperature of the preheated air going to the process and of the beads entering the first packed bed.

In the fluidizing-cooling section

$$(144.3 \tfrac{lbm}{min})(.276 \tfrac{Btu}{lbm\,F})(600 - 100) = (143 \tfrac{lbm}{min})(.292)(1420 - T_{out})$$

$T_{out} = 940°F$ if blower temp = 750, T_{out} = 800°F
                                  appr ΔT = 700°F

(These specific heats are not very accurate.)

This is not a really efficient heat transfer: approach ΔT = 840°F
what if we size the tower for beads entering at 500°F?

Appendix F

Woven Tube and Paper Finned Plate Design Calculations

Hx Design Approach and Cost Development: General

The design approach used involved choosing some performance goals of max ΔP's (hydraulic performance) and minimum required UA (thermal performance) based on previously chosen flow rates, temperatures, and materials/gas properties. These are listed in Tables F-1, F-2, and F-3. Next, tube sizes, spacing, duct height, and width were chosen and analyzed for pressure drops and UA. This geometric model was varied until the pressure drops were below the maximum concurrent with the UA being above the minimum goals. A design close to optimum was reached when the pressure drops were just below the maximum allowed and the UA slightly above the minimum required. Maximum flow conditions were more restrictive than design conditions. A minimum cost Hx design was evaluated based on material volume in the Hx--the assumption being that material volume would drive the cost if the same type of fabrication methods and techniques were used for all designs. Thus, a design meeting the performance goals while having minimum material volume would cost less than an equivalently performing design with more material volume.

Once Hx material volumes were established, costs were generated on a "typical" density and $/lbm conversion basis for each material: SiC, ZrO_2, and Al_2O_3 rigidized fiber products. These basic materials costs (plus any necessary metallic stage) were multiplied by 2.1 to arrive at an initial capital cost of the Hx unit. The 2.1 factor allows for fabrication, processing equipment, and additional low cost materials used as housings, internal ducting, and insulation. It also allows for a suitable safety factor for the metallic stage. Twenty percent of the initial capital cost is taken as the installation cost and 60% of the initial capital cost as the replacement cost (as suggested by Perry: Chemical Engineering Handbook).

Appendix F: Tube and Plate Design Calculations

See Tables F.4-F.11 for the final design parameters which were obtained for the ceramic unit of the ceramic-metal system. See Figures F-1 and F.2 for reference schematics for the design dimensions. See Tables F.12 and F.13 for the design parameters for the all ceramic system.

Woven Tube Concept Design Specifics

Analysis of the woven tube designs is predicated on the following assumptions: the tube to tubesheet transition is well-rounded resulting in entrance and exit flow coefficients of 0.04 and 1.0 respectively (Crane: flow of fluids p. A-29); there is heat transfer across the tubesheets; rectangular ducts are used with round tubes axially oriented along the height dimension (i.e., running top to bottom not side to side) (see Figure F-1); the loss coefficient due to turning (for multiple pass units) is ~3.6 for each turn (Idel'Chik: Handbook of Hydraulic Resistance, diag. 6-21) based on rectangular turning ducts of 1.0 aspect ratio and a between pass length of zero; and a staggered tube arrangement.

Analysis is performed on a HP-41C calculator using a program specifically developed utilizing the assumptions just listed. Initially, geometry parameters, air and gas stream conditions, and air/gas/material properties are input as is the maximum allowed gas ΔP. After calculating the number of tubes in the first row and other assorted flow variables the maximum number of total tube rows is calculated based on gas ΔP max (Kreith: Principles of Heat Transfer, eq. 9-11 and 9-12). The user then chooses the number of air side passes and the total number of tube rows which does not exceed the maximum number. The only restriction is that an integer value will result when dividing the total number of rows by the number of passes. The gas side ΔP is then calculated based on the total number of rows. If the gas side Reynolds number is below 1000, the user is referred to Figure 9.18 (Kreith: Principles of Heat Transfer) for the gas side friction factor. The air side ΔP is then calculated by manipulation of Darcy's formula for heat loss. Air side friction factors are calculated by one of four single phase correlations from the TRAC computer code. The applicable correlation is chosen based on Reynolds number and relative roughness, ε/D. Absolute roughness, ε, can be input by the user or for ceramic coated woven fabric defaults to 0.001 inch. fL/D or K_{pipe} is added to the other loss coefficients--K_{ent}, K_{exit}, and K_{turn} (as specified by the geometry) and then K_{total} is input to the manipulated

Darcy formula to get air side pressure drop. At this point the user can modify the geometry and start the analysis over if the pressure drops are not satisfactory or continue if they are.

Continuing, the gas and air heat exchange areas are calculated. If the gas Reynolds number is less than 1000, the user must look up the Colburn-j factor from Figure 9-18 (Kreith--Principles of Heat Transfer), whereupon the gas film coefficient is backed out. For Reynolds numbers greater than 1000 equation 9-10 (Kreith--Principles of Heat Transfer) is used to get the gas film coefficient. Next the beam length is calculated assuming a shape factor of 1.0 and gas emissivity and absorptivity must be input. Since radiation varies as the 4th power of the temperature, a 4th power average is used for average gas temperature in the standard radiation equations 5-66 and 5-67 (Kreith--Principles of Heat Transfer). Unit gas thermal resistance is then the inverse sum of the gas film, fouling factor, and radiation coefficients. Unit thermal conductive resistance is the material thermal conductivity divided by the conduction path length. For air side Reynolds numbers less than 2100, equation 8-28 (Kreith--Principles of Heat Transfer) is used to find the air film coefficients. For Reynolds numbers greater than 2100, equation 8.21 (Kreith--Principles of Heat Transfer) is used. The air side unit thermal resistance based on gas side area is the inverse of the air side film coefficient times the air side/gas side Hx area ratio. The three resistances are added together. The inverse of this sum is the overall unit conductance based on gas side Hx area, U. When multiplied by the gas side Hx area, the output is UA_{calc}. If the answer is satisfactory the material volume of the Hx can be calculated. Otherwise, some or all of the geometry variables can be changed and the calculation rerun.

Although this iterative technique is tedious, certain trends are noticeable. Tall, narrow ducts (w/long tube lengths) have a better UA for a specific pressure drop than do short, wide ducts (w/short tube lengths). Getting as close as possible to both the maximum allowable gas and air ΔP's and the minimum required UA yields minimum volumes. The more compact the heat exchanger, the less the material volume, obviously.

Appendix F: Tube and Plate Design Calculations   203

Paper Hx Concept Design Specifics

Analysis of the paper Hx designs is predicated on the following assumptions:

Air side entrance and exit venturis are straight lined and single angled and are the same for all passes, there is only one set of entrance and exit venturis on the gas side (constant flow between passes on gas side), there is one more gas passage level than air, there is no heat transfer through the outermost plates, full length fins are used (for each pass), and the loss coefficient due to turning (for multiple pass units) is ~3.6 for each turn (Idel'Chik--Handbook of Hydraulic Resistance, Diag. 6-12) based on rectangular turning ducts of 1.0 aspect ratio and a between pass length of zero (see Figure F.2).

Analysis is performed on a HP-41C calculator utilizing a program specifically developed utilizing the assumptions just listed. Initially, geometry parameters, air and gas stream conditions, and air/gas/material properties are input prior to starting the calculations. Flow areas, Hx areas, wetted perimeters, hydraulic diameters, mass velocities, and Reynolds numbers are calculated for each stream. Currently, friction factors are calculated by Equation 8-18 (Kreith--Principles of Heat Transfer). Surface roughness is ignored. In the interests of greater accuracy or ease, a modification to the program to incorporate the four single phase friction factor correlations from the TRAC computer program is planned. ΔP's for each stream are calculated by manipulation of Darcy's formula for head loss. Entrance and exit loss coefficients, turning loss coefficients (where applicable), and the core loss coefficients are summed prior to use in the ΔP calculations. The results are displayed and the user has the option of changing various geometry parameters and starting the analysis over if the pressure drops are not satisfactory or continuing if they are.

Continuing, the Reynolds numbers and geometry are checked for laminar or turbulent flow and long or short ducts. Calculation of the film coefficient for laminar flow in short ducts uses Equation 8-27 while long ducts use Equation 8-28. Turbulent flow in short ducts uses Equation 8-23 while long ducts use Equation 8-20. (All equations are from Kreith--Principles of Heat Transfer.) Next the radiation coefficient is calculated (using the 4th power average temperature as in the woven tube calculation) using equation 5-66 and 5-67 (Kreith--Principles of Heat Transfer). Unit gas thermal resistivity is then the inverse sum of the gas film, fouling factor, and radiation coefficients. Unit thermal conductive resistance is the material thermal conductivity divided by the conduction path length. The air side unit thermal resistance based on the gas side Hx area is the inverse of the air side film coefficient times the air side/gas side Hx area ratio. The three resistances are then added together; the inverse of this sum is the overall unit conductance based on gas side Hx area, U. When multiplied by the gas side Hx area, the output is UA_{calc}. If the answer is satisfactory, the material volume of the Hx can be calculated. Otherwise, some or all of the geometry variables can be changed and the calculation rerun.

This iterative technique is also quite tedious and similar conclusions to the woven design can be drawn. The more compact the heat exchanger, the less the material volume. Close fin spacings will, of course, foul more easily and quickly--experience is the best instructor here for how close to go on the gas side. The calculations don't take into account conduction into and through the fins. Very close air side fin spacings (especially with low conductivity materials) can cause more pressure drop without sufficient removal of heat to justify it. This trade-off was not studied since relatively large fin spacings were used and substantial calculator program memory was unavailable.

Appendix F: Tube and Plate Design Calculations   205

TABLE F.1. STEEL SOAKING PIT (WITH 2% EXCESS AIR)

CERAMIC PART OF CERAMIC + METAL INSTALLATION

STREAM CONDITIONS & PROPERTIES

		DESIGN	MAXIMUM
AIR $\dot{m}$	LBM/HR	8654	13422
GAS $\dot{m}$	LBM/HR	9114	14130
AIR T_{IN}	°F	1300	1047
GAS T_{IN}	°F	2450	1700
AIR T_{OUT}	°F	2000	1420
GAS T_{OUT}	°F	1887	1395
AIR ρ_{BULK}	LBM/FT³	.0190	.0239
GAS ρ_{BULK}	LBM/FT³	.0152	.0198
AIR μ_{BULK}	LBM/FT·SEC	3.1350×10^{-5}	2.7175×10^{-5}
GAS μ_{BULK}	LBM/FT·SEC	3.5309×10^{-5}	3.0428×10^{-5}
AIR k_{BULK}	BTU/HR·FT·°F	.0421	.0357
GAS k_{BULK}	BTU/HR·FT·°F	.0484	.0401
MAT'L k	BTU/HR·FT·°F	≃10	≃10
AIR $C_{p\,BULK}$	BTU/LBM·°F	.2834	.2743
GAS $C_{p\,BULK}$	BTU/LBM·°F	.3320	.3180

PERFORMANCE PARAMETERS

		DESIGN	MAXIMUM
EFFECTIVENESS, ε		.609	.571
# OF AIR SIDE PASSES		~16	~14
# OF TRANSFER UNITS, N_{tu}			
MINIMUM UA REQUIRED	BTU/HR·°F	3924	5154
MAXIMUM GAS ΔP	INCHES W.C.	7.5	
MAXIMUM AIR ΔP	INCHES W.C.	25.0	

TABLE F2. ALUMINUM REMELT (WITH 2% EXCESS AIR)   CERAMIC PART OF CERAMIC + METAL INSTALLATION

STREAM CONDITIONS & PROPERTIES

AIR $\dot{m}$	LBM/HR	9990
GAS $\dot{m}$	LBM/HR	10518
AIR T_{IN}	°F	951
GAS T_{IN}	°F	2100
AIR T_{OUT}	°F	1800
GAS T_{OUT}	°F	1409
AIR ρ_{BULK}	LBM/FT^3	.0219
GAS ρ_{BULK}	LBM/FT^3	.0181
AIR μ_{BULK}	LBM/FT·SEC	2.8680×10^{-5}
GAS μ_{BULK}	LBM/FT·SEC	3.2291×10^{-5}
AIR k_{BULK}	BTU/HR·FT·°F	.0380
GAS k_{BULK}	BTU/HR·FT·°F	.0436
MAT'L k	BTU/HR·FT·°F	≈10
AIR $C_{P\,BULK}$	BTU/LBM·°F	.2777
GAS $C_{P\,BULK}$	BTU/LBM·°F	.3228

PERFORMANCE PARAMETERS

EFFECTIVENESS, ϵ		.74
# OF AIR SIDE PASSES		2
# OF TRANSFER UNITS, N_{tu}		~3.5
MINIMUM $UA_{REQUIRED}$	BTU/HR·°F	9710
MAXIMUM GAS ΔP	INCHES W.C.	7.5
MAXIMUM AIR ΔP	INCHES W.C.	25.0

TABLE F.3. GLASS MELTING (WITH 2% EXCESS AIR) — ALL CERAMIC UNIT

STREAM CONDITIONS & PROPERTIES

AIR $\dot{m}$	LBM/HR	54000
GAS $\dot{m}$	LBM/HR	57000
AIR T_{IN}	°F	100
GAS T_{IN}	°F	2800
AIR T_{OUT}	°F	2000
GAS T_{OUT}	°F	1365
AIR ρ_{BULK}	LBM/FT³	.0264
GAS ρ_{BULK}	LBM/FT³	.0156
AIR μ_{BULK}	LBM/FT·SEC	2.5230×10^{-5}
GAS μ_{BULK}	LBM/FT·SEC	3.4296×10^{-5}
AIR k_{BULK}	BTU/HR·FT·°F	.0327
GAS k_{BULK}	BTU/HR·FT·°F	.0477
MAT'L k	BTU/HR·FT·°F	≈ 10
AIR $C_{p\,BULK}$	BTU/LBM·°F	.2694
GAS $C_{p\,BULK}$	BTU/LBM·°F	.3205

PERFORMANCE PARAMETERS

EFFECTIVENESS, ε		.704
# OF AIR SIDE PASSES		3
# OF TRANSFER UNITS, N_{tu}		~2.5
MINIMUM $UA_{REQUIRED}$	BTU/HR·°F	36369
MAXIMUM GAS ΔP	INCHES W.C.	8.0
MAXIMUM AIR ΔP	INCHES W.C.	30.0

TABLE F-4.
STEEL SOAKING PIT — PART METAL-PART CERAMIC (2 PASS) SiC CERAMIC UNIT

DESIGN REQUIREMENTS		DESIGN	MAXIMUM
MAXIMUM GAS ΔP	INCHES W.C.	7.5	7.5
MAXIMUM AIR ΔP	INCHES W.C.	25.0	25.0
MINIMUM UA required	BTU/HR·°F	3924	5154
FOULING FACTOR		0	0

WOVEN TUBE HX DESIGN: GEOMETRY & PERFORMANCE & COSTS

DUCT HEIGHT × WIDTH	INCH × INCH	48 × 20	
TUBESHEET LENGTH / PASS	INCH	24.75	
TUBE O.D. × WALL	INCH × INCH	.875 × .040	
STAGGERED TUBE PITCH	INCH	1.375	
# OF TUBES / PASS		280	
GAS SIDE Hx AREA	FT²	525.32	
MATERIAL VOLUME	FT³	1.6625	
GAS ΔP calc	INCHES W.C.	4.03	6.77
AIR ΔP calc	INCHES W.C.	9.92	17.69
U calc (on gas side area)	BTU/HR·FT²·°F	8.29	9.96
UA calc	BTU/HR·°F	4355	5232
INITIAL CAPITAL COST	$	176.8 K	
MAINTENANCE COST	$	35.4 K	
REPLACEMENT COST	$	106.1 K	

PAPER HX CONCEPT DESIGN: GEOMETRY & PERFORMANCE & COSTS

# OF GAS PASSAGE LEVELS		34	
PLATE SPACING, AIR × GAS	INCH × INCH	.25 × .375	
PLATE THICKNESS	INCH	.020	
GAS FLOW W – AIR FLOW L	INCH	18	
GAS FLOW L – AIR FLOW W	INCH	18	
FIN L / INCH W, AIR × GAS	IN./IN. × IN./IN.	1.50 × 1.33	
FIN CONTACT %, AIR × GAS	% × %	30 × 30	
FIN THICKNESS	INCH	.020	
ENTRANCE-EXIT VENTURI ℓ, θ	°	30	
GAS SIDE Hx AREA	FT²	530.39	
MATERIAL VOLUME	FT³	1.1800	
GAS ΔP calc	INCHES W.C.	2.46	3.86
AIR ΔP calc	INCHES W.C.	8.96	17.33
U calc (on gas side area)	BTU/HR·FT²·°F	8.82	9.71
UA calc	BTU/HR·°F	4677	5151
INITIAL CAPITAL COST	$	176.9 K	
MAINTENANCE COST	$	35.4 K	
REPLACEMENT COST	$	106.1 K	

Appendix F: Tube and Plate Design Calculations   209

TABLE F-5
STEEL SOAKING PIT — PART METAL-PART CERAMIC (2 PASS) SiC CERAMIC UNIT

DESIGN REQUIREMENTS   MAXIMUM CONDITIONS

MAXIMUM GAS ΔP	INCHES W.C.	7.5
MAXIMUM AIR ΔP	INCHES W.C.	25.0
MINIMUM UA required	BTU/HR·°F	5154
FOULING FACTOR		.003

WOVEN TUBE HX DESIGN: GEOMETRY & PERFORMANCE & COSTS

DUCT HEIGHT × WIDTH	INCH × INCH	48 × 20
TUBESHEET LENGTH / PASS	INCH	25.94
TUBE O.D. × WALL	INCH × INCH	.875 × .040
STAGGERED TUBE PITCH	INCH	1.375
# OF TUBES / PASS		294
GAS SIDE Hx AREA	FT^2	551.56
MATERIAL VOLUME	FT^3	1.7455
GAS ΔP calc	INCHES W.C.	7.10
AIR ΔP calc	INCHES W.C.	16.13
U calc (on gas side area)	BTU/HR·FT^2·°F	9.42
UA calc	BTU/HR·°F	5196
INITIAL CAPITAL COST	$	184.1 K
MAINTENANCE COST	$	36.8 K
REPLACEMENT COST	$	110.5 K

PAPER HX CONCEPT DESIGN: GEOMETRY & PERFORMANCE & COSTS

# OF GAS PASSAGE LEVELS		
PLATE SPACING, AIR × GAS	INCH × INCH	
PLATE THICKNESS	INCH	
GAS FLOW W - AIR FLOW L (pass)	INCH	
GAS FLOW L - AIR FLOW W (pass)	INCH	
FIN L / INCH W, AIR × GAS	IN./IN. × IN./IN.	
FIN CONTACT %, AIR × GAS	% × %	
FIN THICKNESS	INCH	
ENTRANCE-EXIT VENTURI ≼, θ	°	
GAS SIDE Hx AREA	FT^2	
MATERIAL VOLUME	FT^3	
GAS ΔP calc	INCHES W.C.	
AIR ΔP calc	INCHES W.C.	
U calc (on gas side area)	BTU/HR·FT^2·°F	
UA calc	BTU/HR·°F	
INITIAL CAPITAL COST	$	
MAINTENANCE COST	$	
REPLACEMENT COST	$	

TABLE F-6.
STEEL SOAKING PIT — PART METAL-PART CERAMIC (2 PASS) SiC CERAMIC UNIT

DESIGN REQUIREMENTS

		MAXIMUM CONDITIONS
MAXIMUM GAS ΔP	INCHES W.C.	7.5
MAXIMUM AIR ΔP	INCHES W.C.	25.0
MINIMUM UA required	BTU/HR·°F	5154
FOULING FACTOR		.01

WOVEN TUBE HX DESIGN: GEOMETRY & PERFORMANCE & COSTS

DUCT HEIGHT × WIDTH	INCH × INCH	48 × 22
TUBESHEET LENGTH / PASS	INCH	21.32
TUBE O.D. × WALL	INCH × INCH	.750 × .040
STAGGERED TUBE PITCH	INCH	1.25
# OF TUBES / PASS		350
GAS SIDE Hx AREA	FT^2	561.37
MATERIAL VOLUME	FT^3	1.7640
GAS ΔP_{calc}	INCHES W.C.	5.42
AIR ΔP_{calc}	INCHES W.C.	24.27
U calc (on gas side area)	BTU/HR·FT^2·°F	9.74
UA calc	BTU/HR·°F	5466
INITIAL CAPITAL COST	$	1857 K
MAINTENANCE COST	$	37.1 K
REPLACEMENT COST	$	111.4 K

PAPER HX CONCEPT DESIGN: GEOMETRY & PERFORMANCE & COSTS

# OF GAS PASSAGE LEVELS		
PLATE SPACING, AIR × GAS	INCH × INCH	
PLATE THICKNESS	INCH	
GAS FLOW W - AIR FLOW L (ALT.)	INCH	
GAS FLOW L - AIR FLOW W (ALT.)	INCH	
FIN L / INCH W, AIR × GAS	IN./IN. × IN./IN.	
FIN CONTACT %, AIR × GAS	% × %	
FIN THICKNESS	INCH	
ENTRANCE-EXIT VENTURI ⋖, θ	°	
GAS SIDE Hx AREA	FT^2	
MATERIAL VOLUME	FT^3	
GAS ΔP_{calc}	INCHES W.C.	
AIR ΔP_{calc}	INCHES W.C.	
U calc (on gas side area)	BTU/HR·FT^2·°F	
UA calc	BTU/HR·°F	
INITIAL CAPITAL COST	$	
MAINTENANCE COST	$	
REPLACEMENT COST	$	

Appendix F: Tube and Plate Design Calculations   211

TABLE F-7.
STEEL SOAKING PIT PART METAL-PART CERAMIC (2 PASS) SiC CERAMIC UNIT

DESIGN REQUIREMENTS

		MAXIMUM CONDITIONS
MAXIMUM GAS ΔP	INCHES W.C.	7.5
MAXIMUM AIR ΔP	INCHES W.C.	25.0
MINIMUM UA REQUIRED	BTU/HR·°F	5154
FOULING FACTOR		.05

WOVEN TUBE HX DESIGN: GEOMETRY & PERFORMANCE & COSTS

DUCT HEIGHT × WIDTH	INCH × INCH	48 × 20
TUBESHEET LENGTH / PASS	INCH	29.88
TUBE O.D. × WALL	INCH × INCH	.625 × .040
STAGGERED TUBE PITCH	INCH	1.10
# OF TUBES / PASS		558
GAS SIDE Hx AREA	FT^2	745.4
MATERIAL VOLUME	FT^3	2.3184
GAS ΔP_{calc}	INCHES W.C.	7.49
AIR ΔP_{calc}	INCHES W.C.	24.96
U CALC (ON GAS SIDE AREA)	BTU/HR·FT^2·°F	7.15
UA CALC	BTU/HR·°F	5327
INITIAL CAPITAL COST	$	234.2 K
MAINTENANCE COST	$	46.8 K
REPLACEMENT COST	$	140.5 K

PAPER HX CONCEPT DESIGN: GEOMETRY & PERFORMANCE & COSTS

# OF GAS PASSAGE LEVELS		
PLATE SPACING, AIR × GAS	INCH × INCH	
PLATE THICKNESS	INCH	
GAS FLOW W - AIR FLOW L (≈)	INCH	
GAS FLOW L - AIR FLOW W (≈)	INCH	
FIN L / INCH W, AIR × GAS	IN./IN. × IN./IN.	
FIN CONTACT %, AIR × GAS	% × %	
FIN THICKNESS	INCH	
ENTRANCE - EXIT VENTURI ∢, θ	°	
GAS SIDE Hx AREA	FT^2	
MATERIAL VOLUME	FT^3	
GAS ΔP_{calc}	INCHES W.C.	
AIR ΔP_{calc}	INCHES W.C.	
U CALC (ON GAS SIDE AREA)	BTU/HR·FT^2·°F	
UA CALC	BTU/HR·°F	
INITIAL CAPITAL COST	$	
MAINTENANCE COST	$	
REPLACEMENT COST	$	

TABLE F8
STEEL SOAKING PIT — PART METAL-PART CERAMIC (2 PASS)   SiC CERAMIC UNIT

DESIGN REQUIREMENTS

		DESIGN	MAXIMUM
MAXIMUM GAS ΔP	INCHES W.C.	3.75	3.75
MAXIMUM AIR ΔP	INCHES W.C.	12.5	12.5
MINIMUM UA required	BTU/HR·°F	3924	5154
FOULING FACTOR		0	

WOVEN TUBE HX DESIGN: GEOMETRY & PERFORMANCE & COSTS

DUCT HEIGHT × WIDTH	INCH × INCH	48 × 30	
TUBESHEET LENGTH / PASS	INCH	17.16	
TUBE O.D. × WALL	INCH × INCH	.875 × .040	
STAGGERED TUBE PITCH	INCH	1.30	
# OF TUBES / PASS		338	
GAS SIDE Hx AREA	FT²	631.84	
MATERIAL VOLUME	FT³	1.9992	
GAS ΔP_{calc}	INCHES W.C.	2.18	3.68
AIR ΔP_{calc}	INCHES W.C.	6.99	12.40
U calc (on gas side area)	BTU/HR·FT²·°F	7.13	8.37
UA calc	BTU/HR·°F	4505	5289
INITIAL CAPITAL COST	$	206.3K	
MAINTENANCE COST	$	41.3K	
REPLACEMENT COST	$	123.8K	

PAPER HX CONCEPT DESIGN: GEOMETRY & PERFORMANCE & COSTS

# OF GAS PASSAGE LEVELS		
PLATE SPACING, AIR × GAS	INCH × INCH	
PLATE THICKNESS	INCH	
GAS FLOW W – AIR FLOW L (≡)	INCH	
GAS FLOW L – AIR FLOW W (≡)	INCH	
FIN L / INCH W, AIR × GAS	IN./IN. × IN./IN.	
FIN CONTACT %, AIR × GAS	% × %	
FIN THICKNESS	INCH	
ENTRANCE-EXIT VENTURI ⊄, θ	•	
GAS SIDE Hx AREA	FT²	
MATERIAL VOLUME	FT³	
GAS ΔP_{calc}	INCHES W.C.	
AIR ΔP_{calc}	INCHES W.C.	
U calc (on gas side area)	BTU/HR·FT²·°F	
UA calc	BTU/HR·°F	
INITIAL CAPITAL COST	$	
MAINTENANCE COST	$	
REPLACEMENT COST	$	

Appendix F: Tube and Plate Design Calculations   213

TABLE F-9.
STEEL SOAKING PIT PART METAL-PART CERAMIC (2 PASS) SiC CERAMIC UNIT

DESIGN REQUIREMENTS

		DESIGN	MAXIMUM
MAXIMUM GAS ΔP	INCHES W.C.	11.25	11.25
MAXIMUM AIR ΔP	INCHES W.C.	37.50	37.50
MINIMUM UA required	BTU/HR·°F	3924	5154
FOULING FACTOR			0

WOVEN TUBE HX DESIGN: GEOMETRY & PERFORMANCE & COSTS

DUCT HEIGHT × WIDTH	INCH × INCH	44 × 18	
TUBESHEET LENGTH / PASS	INCH	22.32	
TUBE O.D. × WALL	INCH × INCH	.750 × .040	
STAGGERED TUBE PITCH	INCH	1.25	
# OF TUBES / PASS		280	
GAS SIDE Hx AREA	FT²	413.18	
MATERIAL VOLUME	FT³	1.7980	
GAS ΔP_{calc}	INCHES W.C.	4.68	7.88
AIR ΔP_{calc}	INCHES W.C.	19.97	35.60
U calc (on gas side area)	BTU/HR·FT²·°F	10.67	12.82
UA calc	BTU/HR·°F	4409	5296
INITIAL CAPITAL COST	$	145.0 K	
MAINTENANCE COST	$	29.0 K	
REPLACEMENT COST	$	87.0 K	

PAPER HX CONCEPT DESIGN: GEOMETRY & PERFORMANCE & COSTS

# OF GAS PASSAGE LEVELS		
PLATE SPACING, AIR × GAS	INCH × INCH	
PLATE THICKNESS	INCH	
GAS FLOW W - AIR FLOW L (☐☐)	INCH	
GAS FLOW L - AIR FLOW W (☐☐)	INCH	
FIN L / INCH W, AIR × GAS	IN./IN. × IN./IN.	
FIN CONTACT %, AIR × GAS	% × %	
FIN THICKNESS	INCH	
ENTRANCE·EXIT VENTURI ∢, θ	°	
GAS SIDE Hx AREA	FT²	
MATERIAL VOLUME	FT³	
GAS ΔP_{calc}	INCHES W.C.	
AIR ΔP_{calc}	INCHES W.C.	
U calc (on gas side area)	BTU/HR·FT²·°F	
UA calc	BTU/HR·°F	
INITIAL CAPITAL COST	$	
MAINTENANCE COST	$	
REPLACEMENT COST	$	

TABLE F-10.
ALUMINUM REMELT PART CERAMIC-PART METAL (2 PASS) ZrO₂ CERAMIC UNIT

DESIGN REQUIREMENTS

MAXIMUM GAS ΔP	INCHES W.C.	7.5
MAXIMUM AIR ΔP	INCHES W.C.	25.0
MINIMUM UA required	BTU/HR·°F	9710
FOULING FACTOR		0

WOVEN TUBE HX DESIGN: GEOMETRY & PERFORMANCE & COSTS

DUCT HEIGHT × WIDTH	INCH × INCH	48 × 21
TUBESHEET LENGTH/PASS	INCH	33.14
TUBE O.D. × WALL	INCH × INCH	.500 × .040
STAGGERED TUBE PITCH	INCH	.95
# OF TUBES / PASS		880
GAS SIDE Hx AREA	FT²	939.26
MATERIAL VOLUME	FT³	2.8745
GAS ΔP_{calc}	INCHES W.C.	4.58
AIR ΔP_{calc}	INCHES W.C.	21.89
U calc (on gas side area)	BTU/HR·FT²·°F	10.38
UA calc	BTU/HR·°F	9752
INITIAL CAPITAL COST	$	205.3 K
MAINTENANCE COST	$	41.1 K
REPLACEMENT COST	$	123.2 K

PAPER HX CONCEPT DESIGN: GEOMETRY & PERFORMANCE & COSTS

# OF GAS PASSAGE LEVELS		36
PLATE SPACING, AIR × GAS	INCH × INCH	.1875 × .250
PLATE THICKNESS	INCH	.020
GAS FLOW W - AIR FLOW L (FREE)	INCH	24
GAS FLOW L - AIR FLOW W (FREE)	INCH	24
FIN L / INCH W, AIR × GAS	IN./IN. × IN./IN.	1.50 × 1.33
FIN CONTACT %, AIR × GAS	% × %	30 × 30
FIN THICKNESS	INCH	.020
ENTRANCE-EXIT VENTURI ⦝, θ	°	30
GAS SIDE Hx AREA	FT²	998.39
MATERIAL VOLUME	FT³	2.2222
GAS ΔP_{calc}	INCHES W.C.	7.32
AIR ΔP_{calc}	INCHES W.C.	15.07
U calc (on gas side area)	BTU/HR·FT²·°F	9.76
UA calc	BTU/HR·°F	9740
INITIAL CAPITAL COST	$	165.9 K
MAINTENANCE COST	$	33.2 K
REPLACEMENT COST	$	99.5 K

Appendix F: Tube and Plate Design Calculations   215

TABLE F-11.
GLASS MELTING   ALL CERAMIC (3 PASS) Al_2O_3 CERAMIC UNIT

DESIGN REQUIREMENTS

MAXIMUM GAS ΔP	INCHES W.C.	8.0
MAXIMUM AIR ΔP	INCHES W.C.	30.0
MINIMUM UA REQUIRED	BTU/HR·°F	36369
FOULING FACTOR		0

WOVEN TUBE HX DESIGN: GEOMETRY & PERFORMANCE & COSTS

DUCT HEIGHT × WIDTH	INCH × INCH	60 × 62
TUBESHEET LENGTH/PASS	INCH	72.55
TUBE O.D. × WALL	INCH × INCH	.875 × .040
STAGGERED TUBE PITCH	INCH	2.15
# OF TUBES / PASS		1112
GAS SIDE Hx AREA	FT^2	3999.11
MATERIAL VOLUME	FT^3	12.6862
GAS ΔP CALC	INCHES W.C.	7.73
AIR ΔP CALC	INCHES W.C.	29.95
U CALC (ON GAS SIDE AREA)	BTU/HR·FT^2·°F	10.07
UA CALC	BTU/HR·°F	40290
INITIAL CAPITAL COST	$	648.2K
MAINTENANCE COST	$	129.2K
REPLACEMENT COST	$	388.9K

PAPER HX CONCEPT DESIGN: GEOMETRY & PERFORMANCE & COSTS

# OF GAS PASSAGE LEVELS		64
PLATE SPACING, AIR × GAS	INCH × INCH	.260 × .570
PLATE THICKNESS	INCH	.020
GAS FLOW W - AIR FLOW L (FT)	INCH	24
GAS FLOW L - AIR FLOW W (FT)	INCH	24
FIN L / INCH W, AIR × GAS	IN./IN. × IN./IN.	1.50 × 1.33
FIN CONTACT %, AIR × GAS	% × %	30 × 30
FIN THICKNESS	INCH	.020
ENTRANCE - EXIT VENTURI ∢, θ	°	30
GAS SIDE Hx AREA	FT^2	2662.37
MATERIAL VOLUME	FT^3	5.9466
GAS ΔP CALC	INCHES W.C.	7.27
AIR ΔP CALC	INCHES W.C.	27.43
U CALC (ON GAS SIDE AREA)	BTU/HR·FT^2·°F	14.04
UA CALC	BTU/HR·°F	37388
INITIAL CAPITAL COST	$	303.8K
MAINTENANCE COST	$	60.8K
REPLACEMENT COST	$	182.3K

TABLE F-12. STEEL SOAKING PIT (WITH 2% EXCESS AIR)   ALL CERAMIC UNIT

STREAM CONDITIONS & PROPERTIES

		DESIGN	MAXIMUM
AIR $\dot{m}$	LBM/HR	8654	13422
GAS $\dot{m}$	LBM/HR	9114	14130
AIR T_{in}	°F	100	100
GAS T_{in}	°F	2450	2700
AIR T_{out}	°F	2000	1420
GAS T_{out}	°F	906	612
AIR ρ_{BULK}	LBM/FT³	.0264	.0325
GAS ρ_{BULK}	LBM/FT³	.0183	.0249
AIR μ_{BULK}	LBM/FT·SEC	2.5230×10^{-5}	2.2060×10^{-5}
GAS μ_{BULK}	LBM/FT·SEC	3.1600×10^{-5}	2.6353×10^{-5}
AIR k_{BULK}	BTU/HR·FT·°F	.0327	.0279
GAS k_{BULK}	BTU/HR·FT·°F	.0425	.0344
MAT'L k	BTU/HR·FT·°F	≈10	≈10
AIR $C_{p\,BULK}$	BTU/LBM·°F	.2694	.2600
GAS $C_{p\,BULK}$	BTU/LBM·°F	.3147	.2996

PERFORMANCE PARAMETERS

		DESIGN	MAXIMUM
EFFECTIVENESS, ε		.809	.825
# OF AIR SIDE PASSES		~52	3
# OF TRANSFER UNITS, N_{tu}		12123	~5.7
MINIMUM UA REQUIRED	BTU/HR·°F		19891
MAXIMUM GAS ΔP	INCHES W.C.		8.0
MAXIMUM AIR ΔP	INCHES W.C.		30.0

Appendix F: Tube and Plate Design Calculations   217

TABLE F13
STEEL SOAKING PIT   (3 PASS)   ALL CERAMIC SiC UNIT

DESIGN REQUIREMENTS

		DESIGN	MAXIMUM
MAXIMUM GAS ΔP	INCHES W.C.	8.0	
MAXIMUM AIR ΔP	INCHES W.C.	30.0	
MINIMUM UA required	BTU/HR·°F	12123	19891
FOULING FACTOR		0.0	

WOVEN TUBE HX DESIGN: GEOMETRY & PERFORMANCE & COSTS

DUCT HEIGHT × WIDTH	INCH × INCH	48 × 22	
TUBESHEET LENGTH/PASS	INCH	65.88	
TUBE O.D. × WALL	INCH × INCH	.450 × .080	
STAGGERED TUBE PITCH	INCH	1.10	
# OF TUBES / PASS		1380	
GAS SIDE Hx AREA	FT^2	2008.27	
MATERIAL VOLUME	FT^3	6.0105	
GAS ΔP_{calc}	INCHES W.C.	4.38	7.18
AIR ΔP_{calc}	INCHES W.C.	17.18	30.07
U $_{calc}$ (ON GAS SIDE AREA)	BTU/HR·FT^2·°F	7.83	9.93
UA $_{calc}$	BTU/HR·°F	15733	19939
INITIAL CAPITAL COST	$	525.5K	
MAINTENANCE COST	$	105.1K	
REPLACEMENT COST	$	315.3K	

PAPER HX CONCEPT DESIGN: GEOMETRY & PERFORMANCE & COSTS

# OF GAS PASSAGE LEVELS		
PLATE SPACING, AIR × GAS	INCH × INCH	
PLATE THICKNESS	INCH	
GAS FLOW W - AIR FLOW L	INCH	
GAS FLOW L - AIR FLOW W	INCH	
FIN L / INCH W, AIR × GAS	IN./IN. × IN./IN.	
FIN CONTACT %, AIR × GAS	% × %	
FIN THICKNESS	INCH	
ENTRANCE - EXIT VENTURI ∡, θ	°	
GAS SIDE Hx AREA	FT^2	
MATERIAL VOLUME	FT^3	
GAS ΔP_{calc}	INCHES W.C.	
AIR ΔP_{calc}	INCHES W.C.	
U $_{calc}$ (ON GAS SIDE AREA)	BTU/HR·FT^2·°F	
UA $_{calc}$	BTU/HR·°F	
INITIAL CAPITAL COST	$	
MAINTENANCE COST	$	
REPLACEMENT COST	$	

Appendix G

State-of-the-Art and Emerging Concepts Costs

This appendix provides the cost information details which were used for the state-of-the-art and emerging technology designs for the economic analyses.

For the economics study, it was decided to assume that recuperator capital costs would vary directly as the air flow rate. This is equivalent to saying that modules exist and more or less are placed in parallel to do the particular job. No attempt was made to optimize the units thermally or economically. See Tables 1 through 3 for the cost data for the steel soaking pit, aluminum melting and glass remelting applications respectively.

Appendix G: State-of-the-Art and Emerging Concepts Costs 219

TABLE I: RECUPERATOR COSTS FOR STEEL SOAKING PIT FURNACES
(22 mmBtu/hr Application)

Program/Concept	As-Designed Flow Rate (lbm/hr)	Required Flow Rate (lbm/hr)[1]	Scaling Factor	Initial Capital Costs	Annual Operating Costs[2]	Annual Maintenance Costs[3]	Capital Replacement Costs[4]
LANL Heat Pipe (w/SiC)	19,605	8,285	0.42	$133.9K[5]	$10.4K	$26.8K	$80.3K
Garrett-Airesearch Tube in Shell	8,654/2,886	8,654	---	$250.9K[6]	$10.4K	$50.2K	$150.5K
B & W Tube in Tube	11,250/ ---	8,428	0.75	$345.3K[7]	$10.4K	$69.1K	$207.2K
Solar Turbines Tube in Shell	2,340/ ---	8,346	3.57	$263.2K[8]	$10.4K	$52.6K	$157.9K

NOTES:

1. Provided in Section 4.3.3.
2. Annual operating labor costs: 10 hrs/wk × 52 wks × $20/hr = $10.4K.
3. Assumed to be 20% of capital costs.
4. Assumed to be 60% of capital costs.
5. The initial capital cost for the LANL ceramic heat pipe was developed from information contained in a LASL progress report, LA-8403-PR dated June 1980. The cost summary from the report is as follows:

Ceramic Heat Pipe	$53.0K	Insulation	$3.1K
Ceramic End Support and Central Baffle	$6.8K	Support Structure	$13.7K
		Fans	$9.6K
Metallic Recuperator	$8.1K	Additional Items	$56.0K
Assembly Labor		SUBTOTAL	$160.3K
Heat Pipes	$1.6K		
Metallic Recuperator	$0.7K		
Transition Ducting	$7.7K		

To this subtotal, an additional 35% was added for "profit". This resulted in a total cost of $216.4K (in 1979 $). The cost was then escalated for 4 years at 10% and then multiplied by the scaling factor; i.e., $216.4K × (1.1)^4 × 0.42 ≈ $133.9K.

TABLE I: (continued)

6. The initial capital cost of the Garrett-Airesearch recuperator was determined by modifying Garrett's costs for the Lone Star furnace installation. These costs are identified in Garrett-Airesearch Report GRI-82/0015 (Table 5-13) and are modified as follows:

	Garrett Costs	Escalated & Adjusted Costs
Ceramic Tubing at $15/ft	$34,884	$ 42,210
Ceramic Header Plates, Two at $0.26/cu. in.	500	605
Ceramic Plates for Headers and Sideplates at $0.26/cu. in.	3,270	3,957
Insulation at $5/sq ft for 1-1/2 in. Thickness	2,504	3,030
Steel Casing at $20/sq ft of Steel Plate	2,865	3,467
Steel Support Framing at $20/cu. ft of Recuperator Volume	1,811	2,191
Assembly Labor at $10/tube	5,700	15,000 *
Miscellaneous Parts, Ducts, Controls, and Labor at $10,000 (lb/sec) Airflow	24,050	29,100
Exhaust Fan at $500/Air hp	1,000	4,840
Metallic Heat Exchanger at $12/lb	4,104	5,000
Burners	9,445	11,429
Subtotal (1981 Dollars)	$90,133	$120,829 Escalated & Adjusted Subtotal
		65,000 Installation
		65,040 G&A Profit(35%)
		$250,869 TOTAL

* This cost item adjusted from Garrett-Airesearch estimate.

TABLE I: (continued)

7. The B&W costs were extracted from the B&W prepared cost estimate, dated March 24, 1983, and adjusted. The following is the cost breakdown:

Ceramic Tubes	$ 125.0K
Ceramic Sleeves and Seals	20.0K
Insulation	15.0K
Steel	9.0K
Fabrication	20.0K *
Miscellaneous Parts	47.0K
Fans	25.0K
Metallic Recuperator	5.0K *
Burners	10.0K
Installation	65.0K
G&A Profit (35%)	119.3K
TOTAL	$460.4K

   * These cost items are adjusted from B&W's estimate.

   The total cost of $460.4K was then scaled by the 0.75 scaling factor giving a cost of $345.3K.

8. The Solar Turbines recuperator cost is based on a test module capable of 2340 lbm/hr air flow. The estimated cost of this module is as follows:

Ceramic Tubes @ $10/ft	$ 4,600
Ceramic Header:	$ 5,000
Metal Shell:	$ 2,000
Meal Header:	$ 2,000
Insulation:	$ 2,000
Fan w/Motor:	$ 4,000
Metal Recuperator:	$ 5,000
SUBTOTAL	$24,600
Fabrication & Installation - assumed to be double the equipment costs.	$49,200
TOTAL	$73,800

   Multiplying by the scaling factor gives a total cost equal to $263.2K.

TABLE II: RECUPERATOR COSTS FOR ALUMINUM MELTING FURNACES
(20 mmBtu/hr Application)

Program/Concept	As-Designed Flow Rate (lbm/hr)	Required Flow Rate (lbm/hr)[1]	Scaling Factor	Initial Capital Costs	Annual Operating Costs[2]	Annual Maintenance Costs[3]	Capital Replacement Costs[4]
Garrett-Airesearch	9,990/1,998	9,305	0.93	$338.2K[5]	$10.4K	$67.6K	$202.9K
Uintex/Millcreek Glass	16,200	9,603	0.59	273.8K[6]	10.4K	54.8K	164.3K

NOTES:

1. Provided by C. J. Bliem.
2. Annual operating labor costs: 10 hrs/wk x 52 wks x $20/hr = $10.4K
3. Assumed to be 20% of capital costs.
4. Assumed to be 60% of capital costs.

Continued on next page.

Appendix G: State-of-the-Art and Emerging Concepts Costs   223

TABLE III: RECUPERATOR COSTS FOR GLASS REMELTING FURNACE
(215 mmBtu/hr Application)

Program/Concept	As-Designed Flow Rate (lbm/hr)	Required Flow Rate (lbm/hr)[1]	Scaling Factor	Initital Capital Costs	Annual Operating Costs[2]	Annual Maintenance Costs[3]	Capital Replacement Costs[4]
Uintex/Millcreek	16,200	54,000	4.30[5]	$1,995.7K	$ 10.4K	$ 399.1K	$1,197.4K

NOTES:

1. Provided by C. J. Bliem.
2. Annual operating labor costs: 10 hrs/wk x 52 wks x $20/hr = $10.4K
3. Assumed to be 20% of capital costs.
4. Assumed to be 60% of capital costs.
5. The total scaling factor is developed by multiplying the flow ratio (3.33) by a 29% increase in module stack length; i.e., 3.33 x 1.29 = 4.30.
6. See Table II, Note 6 for as developed capital costs; i.e., $464.1K. Scaled by scaling factor gives the indicated costs.

Part II

Materials Technology

The information in Part II is from *An Assessment of Ceramic Materials Technology for Heat Exchangers* prepared by B.A. Barna, S.P. Henslee, P.V. Kelsey, D.J. Landini, J.C. Mittl, J.F. Whitbeck, D.J. Wiggins and J.M. Zabriskie of EG&G Idaho, Inc. and J.I. Federer and E.S. Bomar of Oak Ridge National Laboratory for the U.S. Department of Energy, January 1984.

1. Introduction

An estimated 1 to 2 quads of energy are lost each year from high temperature, high fouling, corrosive industrial waste streams. This is equivalent to the power produced by 14 to 28 1000-MW power plants. Much of this energy is lost from the aluminum, steel, and glass-making industries. Waste energy from these operations is among the most difficult to recover because the waste energy streams are: (a) high temperature [flue gas temperatures range up to 1650°C (3000°F)]; (b) high fouling (particulates build up on heat exchanger surfaces); (c) corrosive to metallic alloys, especially at desired preheat temperatures of 1100°C (2000°F) or higher. The characteristics of these streams are discussed in more detail elsewhere.[1]

Preheated air temperatures above ~1000°C (~1832°F) preclude use of metallic alloys. Although radiant and convective recuperators successfully use high temperature alloys as partition wall materials (separating flue gases from air), this is accomplished only by limiting the preheated air temperature. The partition wall, therefore, is maintained at a sufficiently low temperature for structural integrity and tolerable oxidation/corrosion. Catastrophic oxidation/corrosion and/or mechanical failure occurs if the partition wall temperature is allowed to exceed limits that are well known for many metallic alloys. Mechanical weakening and/or oxidation limit high temperature Fe-, Ni-, and Co-based alloys to approximately 1000°C (~1830°F). Although coatings (such as aluminized coatings and those consisting of MCrAlY, where M is Fe, Co, or Ni) might be used to improve the oxidation/corrosion resistance of the alloys at temperatures of 1000°C and higher, the mechanical strength of the alloys would not be adequate. Refractory alloys based on W, Mo, Ta, and Nb, which have better high temperature strength than the above alloys, have unacceptable oxidation resistance at the temperatures of interest.

Only ceramic materials have sufficient high temperature strength and oxidation resistance to serve as primary heat exchanger elements providing air preheated to 1100°C. Ceramic materials are being considered for use in other high temperature applications such as in automobile engines and turbochargers. In waste heat recovery systems, ceramics are being formed into tubes or used in granular form in fluidized bed/pebble bed applications.

Optimum design of a waste heat recovery system requires that the characteristics of the material be known. The material must conform to specifications that produce a consistent set of properties. For example, it is well known that many standards, such as ANSI, exist to define steels of a given composition, production method, and heat treatment. As a result, the metals industry has, over the years, developed a comprehensive data base which is available to designers. Moreover, this data base has been formalized through various codes (ASME, for example), standards, and quality assurance methods. Much information is also available on the corrosion of metals in various atmospheres.

Similar information, at even a rudimentary level, may not be readily available in usable form for the ceramic hardware designer. In particular, experience with ceramics in high temperature waste streams appears to be essentially limited to furnace refractories rather than structural ceramics.

The purpose of this report is to assess the data base that exists for structural ceramic materials. This assessment includes the concerns related to the design and use of ceramics for high temperature waste heat applications, including the availability of a variety of ceramic materials, the availability of material properties data, and the applicability of current product reliability techniques.

The following section of this report outlines the current state of ceramic heat exchanger technology to provide a basis for understanding the applications, configurations, and concerns. The next section discusses ceramic recuperator design considerations followed by a section that evaluates various ceramic materials. The next section assesses the technology needed to satisfy design requirements, and the last section presents conclusions and recommendations.

2. Current Ceramic Heat Exchanger Designs

Seven ceramic recuperator concepts that have been or are being developed (Table 1) illustrate the kinds of problems inherent in designing and building ceramic recuperators. Two of these are commercially available and are considered state-of-the-art. The other five are undergoing prototype development and testing and are classed as emerging designs. These seven designs are described in the following paragraphs.

State-of-the-Art Designs

DOE has recognized the need for advanced recuperators and has supported the development and demonstration of two ceramic recuperators which are now available commercially: a finned plate heat exchanger manufactured by GTE Sylvania and a cross counterflow tubular heat exchanger manufactured by Hague International. These ceramic units offer performance advantages over conventional metallic units and are capable of producing 650 to 815°C (1200 to 1500°F) preheated combustion air (i.e., heat exchanger effectiveness up to 50%). However, these recuperators have been applied with minimal success to fouling, high-temperature corrosive streams.

GTE Recuperator

GTE Products Corporation developed a compact, crossflow ceramic recuperator called the "Super Recuper" under a program jointly funded by DOE.[2] The basic crossflow matrix module can operate in a counterflow configuration when staged (Figure 1). The modules are made of extruded plates that are stacked and bonded with a proprietary compound during firing. Various passage sizes are available. Cordierite, a magnesia alumina silicate (MAS), was selected as the principal construction material on the basis of fabricability and its resistance to thermal shock and corrosion. Cordierite softens at about 1400°C (2550°F). The mean coefficient of linear thermal expansion is substantially lower than that of silicon nitride, silicon carbide, mullite, and alumina. The ceramic cubes are available in three sizes designed to transfer approximately 1.8, 2.9, and 4.4 x 10^5 J/s (0.6, 1.0, and 1.5 Btu/h).

TABLE 1. STATE-OF-THE-ART AND EMERGING HEAT EXCHANGER DESIGNS

Company	Hx Type	Materials	Unique Features	Highest Preheat Temperature (°F)	Average Effectiveness (%)	Problems/Uncertainties
State-of-the-Art Designs						
GTE	Finned Plate	Cordierite MAS 8400	Crosswise stacked and bonded; extruded core; spring loaded housing and seals	1500 (calculated)	45 (calculated)	Matrix cracks; MAS softens at 2650°F; leakage of combined modules
Hague	Tube in shell	Oxynitride bonded SiC	Spring loaded mounts and seals; radially finned and smooth tubing	1750	40	Porous tube and seal leakage; pressure drop; large size
Emerging Designs						
Garrett	Tube in shell	SiC tubes Si$_3$N$_4$ bonded SiC tubesheets	Compression loaded structure; cement and compliant seals	2000	75	High air side pressure drop; fouling of metal stage; joint leakage; large size
B&W	Tube in tube	SiC tubes SST tubesheets	Hanging bayonet mount; ceramic paper seals	2000	75	High air side pressure drop; joint leakage; large size
Uintex	Helical	Alumina Chromia Magnesia Chromia	Stackable segments single and double helices with and without fins	1700	50	Casting difficulties; large size and cost
Solar Turbines	Tube in shell	SiC	Viscous glass seals; braze bonded tubes; ball and socket joints	1900	70	Joint leakage; length and cost
Midland-Ross	Heat wheel	MAS	Matrix segments cast in place; resilient bonding to metal housing; labyrinth seals	2000	70	Seal leakage; matrix integrity

Current Ceramic Heat Exchanger Designs 229

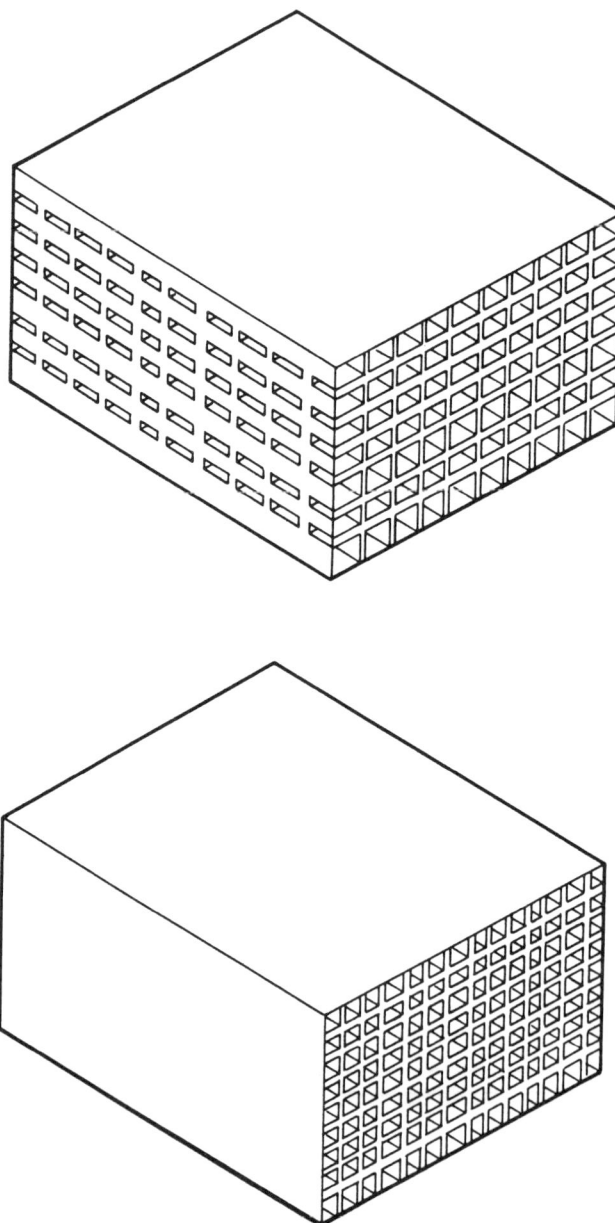

Figure 1. GTE cross-flow recuperator sealed cube.

For the extreme environment of borosilicate glass furnaces, some experiments were made with fluxed zirconia silicate and alumina-coated MAS. Difficulties were encountered in maintaining dimensional tolerances with the fluxed zirconia silicate, and delamination of the ceramic matrix occurred with alumina-coated MAS.

After the development work, DOE and GTE entered into a cooperative agreement titled "Technology Acceleration Program for High Temperature Recuperators" (TAPHTR).[3] Under the terms of this agreement, GTE agreed to install 175 recuperators on various industrial furnaces and measure their performance. The successful completion of this program with demonstrated fuel savings and favorable payback periods would promote widespread industrial acceptance of the GTE recuperator.

Table 2 summarizes the industrial furnace installation under the TAPHTR program, which involved 175 recuperators on 38 furnaces. All furnaces burned natural gas or No. 2 oil. Performance data were collected for 28 of the 38 installations.[4] Unusual operating conditions or problems in collecting data resulted in indicated savings being less than had been estimated in five installations. In the other sites the savings equaled or exceeded estimates. The average total fuel savings from 23 sites was 44%. Part of the savings resulted from improvements to the furnaces when the recuperators were installed.

In a few cases (not identified) operating problems not related to the recuperators caused excessive temperatures, which resulted in destruction of the recuperators; otherwise, no failures occurred. The highest flue gas temperature shown in Table 2 is 1370°C (2500°F). A cordierite recuperator accepting flue gases at this temperature would actually operate at a lower temperature because of flow of cooler combustion air. Such installations, however, require controls that protect the recuperator from softening or melting by shutting down the furnace or admitting dilution air to the flue gases if flow of combustion air stops.

TABLE 2. SUMMARY OF GTE CORPORATION CERAMIC RECUPERATOR INSTALLATIONS UNDER THE TAPHTR

Number of Furnaces	Type of Furnace	Flue Gas Temperature (°C)	Number of Recuperators
13	Box forge	1090-1260	66
3	Slot forge	1205-1370	6
7	Heat treat	925-1260	46
4	Tube furnaces	870-1175	23
2	Die reheat	1095	6
2	Roll reheat	1150-1230	10
2	Aluminum pot melter	1120	2
2	Ladle heater	1095-1205	5
2	Rotary calciner	955-980	6
1	Vacuum report	1205	5
38			175

Results are available for other uses of the cordierite recuperator preceding the TAPHTR program. A recuperator installed on a molybdenum heat treating furnace operated for 2200 h with flue gas inlet temperatures of 1100 to 1400°C (2010 to 2550°F).[5] The recuperator produced air preheated to 800 to 900°C (1470 to 1650°F) with an effectiveness of almost 60% until an accidental overtemperature caused slumping of the cordierite. Although the furnace burned natural gas, product carryover from the furnace (mostly refractory metal oxides) periodically blocked passages. High velocity air was used to blow accumulated material from the recuperator. Ceramographic examination of the recuperator revealed reactions with product carryover; however, failure was mostly attributed to overheating rather than the reactions. The recuperator's performance during this test in a very dirty environment was very encouraging.

Cordierite recuperators on other furnaces have operated successfully. A recuperator on a tungsten reduction furnace operated >4200 h at a flue gas inlet temperature of 925°C (1700°F) and an air outlet temperature of 675°C (1250°F); a recuperator on a rotary calciner operated >4000 h at a flue gas inlet temperature of 760°C (1400°F) and an air outlet temperature of 605°C (1120°F).[6]

Although the Super Recuper performed well in the furnace applications specified in the TAPHTR program, the use of cordierite in the waste streams selected for consideration in this report would be marginal for several reasons:

1. Mechanical weakness as flux gas temperatures approach the softening temperature;

2. Corrosion resistance is poor in borosilicate glass furnace atmosphere, and corrosion data are lacking for aluminum remelt furnace and steel soaking pit atmospheres;

3. Fouling of small passages by flue gas particulates is likely.

More information may be found in References 7 and 8.

Hague International Recuperator

The "Cerhx" recuperator from Hague International, the other commercially available ceramic heat exchanger, is a basic tube-in-shell design using two passes of crossflow tube bundles (Figure 2). The tubes are a proprietary oxynitride bonded silicon carbide (SiC) and are available in plain and finned tube versions. The tubes use spring loaded mounts and seals.[9]

Hague International participated in a program sponsored by the Office of Industrial Programs of DOE to demonstrate energy savings provided by improved steel forging furnaces and heat recovery components. The heat recovery equipment included improved insulation, automatic air-to-fuel ratio control, and the ceramic tube recuperator. Although the recuperator can preheat air to about 1000°C using flue gases at about 1550°C, air preheat temperatures between 600 and 800°C are common in most installations.

Eleven host forge shops were used under the DOE demonstration program.[10] Hague furnaces and heat recovery equipment were installed at these sites. The forge shops furnished performance information on these

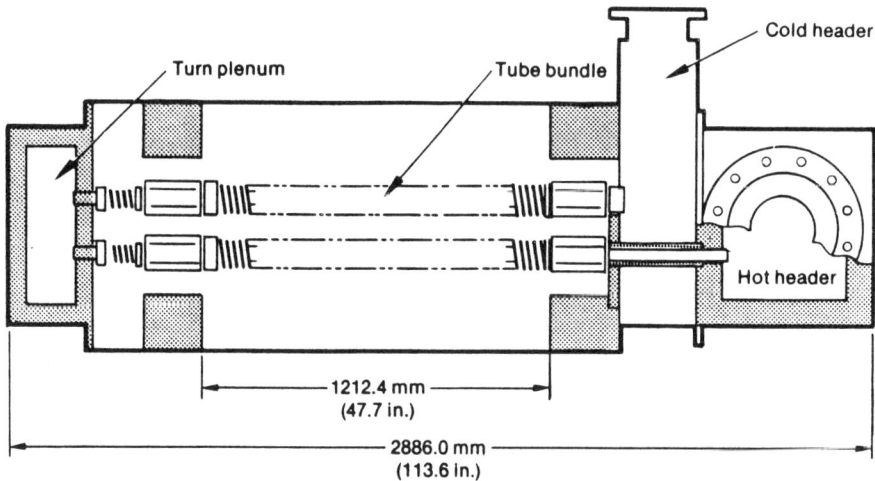

Figure 2. Two-pass ceramic heat exchanger (Hague International).

and other furnaces to a third party for evaluation of energy and economic savings. This information, which basically consisted of cost of installation data and fuel consumption data related to production, was furnished in varying degrees of completeness by the hosts.

The economic analysis revealed fuel savings ranging from 29 to 56% at nine of the eleven host sites. The wide range in fuel savings depended on a number of factors including the condition of the unrecuperated furnace used for comparison and the production practices used in particular sites. At one of the two remaining sites the host reported increased fuel consumption by the Hague furnace compared to an unrecuperated furnace; however, product throughput was not monitored, so this claim is not substantiated. At the the other site a baseline furnace was not available for comparison; however, the specific fuel consumption was the lowest among the host sites.

Materials problems in the furnaces and recuperators are not well documented. During the demonstration program some host sites experienced short lifetimes (undefined) of silicon carbide tubes; however, Hague apparently improved the lifetime of the tubes over the two year period of the program. At the present time, tube life in forging applications averages two years. Problems were also experienced with ceramic materials in burners and furnace linings, but these apparently were solved. Overall, the program was a successful demonstration of improved furnaces and heat recovery equipment.

Cerhx recuperators have also been installed on aluminum remelt furnaces in the U.S. and Japan. Tube life averages one year in these applications. Shorter life compared to forge furnaces is attributed to the corrosive action of halide fluxes used in aluminum remelt furnaces.

Hague International currently manufactures and sells Cerhx recuperators for industrial applications.

Emerging Designs

Under cost-sharing contracts, the Department of Energy (DOE), Garrett-AiResearch Co., and Babcock & Wilcox are developing two hightemperature, burner-duct-recuperator (HTBDR) systems using a technology base that is somewhat advanced compared to existing ceramic heat exchangers. Although the predicted heat exchanger thermal performance of these two recuperators is higher than previous designs (i.e., approximately 80% effectiveness and combustion air preheat to 1090°C or 2000°F), technical risk is high and durability is unknown. The uncertainties are attributable to a very limited design data base for ceramic materials for waste heat recovery applications.

Previous recuperator development activities such as the Uintex/Millcreek glass furnace application have also been supported by DOE. The Gas Research Institute (GRI) has funded several other recuperator projects such as the Solar Turbines International design and the Midland-Ross design. A more detailed discussion of these designs follows.

Garrett-AiResearch Design

The Garrett-AiResearch design is a tube-and-header ceramic recuperator in series with a plate-and-fin metallic recuperator (Figure 3). The ceramic section uses two crossflow passes through SiC tubes. The tubesheets and enclosures are silicon nitride bonded SiC and the assembly is held in place by a system of spring loaded tie rods, square tubes, and cross beam channels constructed mostly of RA330, a high temperature alloy. Difficulties with this design are leakage through the compliant ceramic paper seals at the tube-to-tubesheet joints and high air side pressure drop. This leakage reduces heat exchanger efficiency and the pressure drop may necessitate additional fan capacity. The major unknown for this design's use in the selected waste heat streams is material and joint durability. More information can be found in Reference 11.

Babcock & Wilcox Design

The Babcock & Wilcox HTBDR design features a SiC tube-in-tube configuration (Figure 4). Hot exhaust gas flows on the outside of the hanging tubes. Preheat air flows through the inside of a SiC inner tube and back up the annulus formed by the SiC outer and inner concentric tubes. The inner tube is CS 101, the outer tube is NC 430. Both hang from air cooled, type 309 SST tubesheets. Tube-to-tubesheet seals are compliant ceramic paper. Difficulties with this design are high air side pressure drop, leakage through the compliant seals, and performance of the cooled tubesheet. Again, the major unknown for this design's use in the selected waste heat streams is its durability. More information can be found in Reference 12.

Uintex/Millcreek Design

Glass furnaces present a severe exhaust environment for recuperating waste heat due to the entrainment and carryover of glass batch constituents. Uintex/Millcreek designed and tested a subscale recuperator for use in the harsh environments of a glass furnace. This design is a helical counterflow type using stackable modules (Figure 5). Alumina chromia

Current Ceramic Heat Exchanger Designs   237

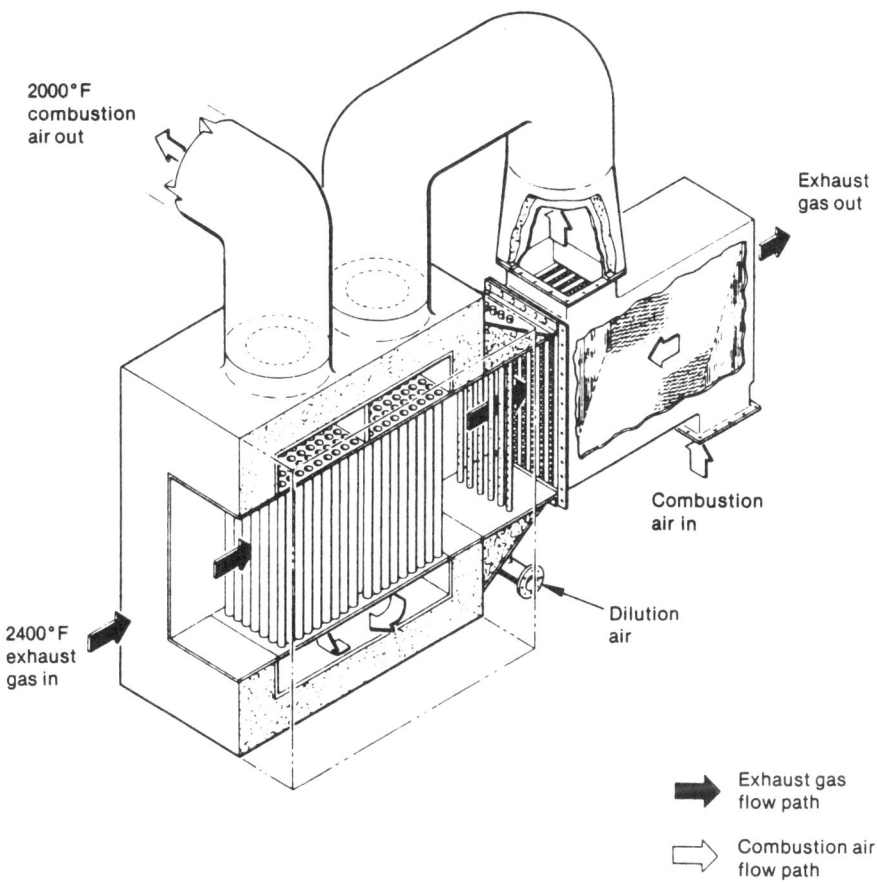

Figure 3. HTBDR ceramic heat exchanger under development by Garrett-AiResearch.

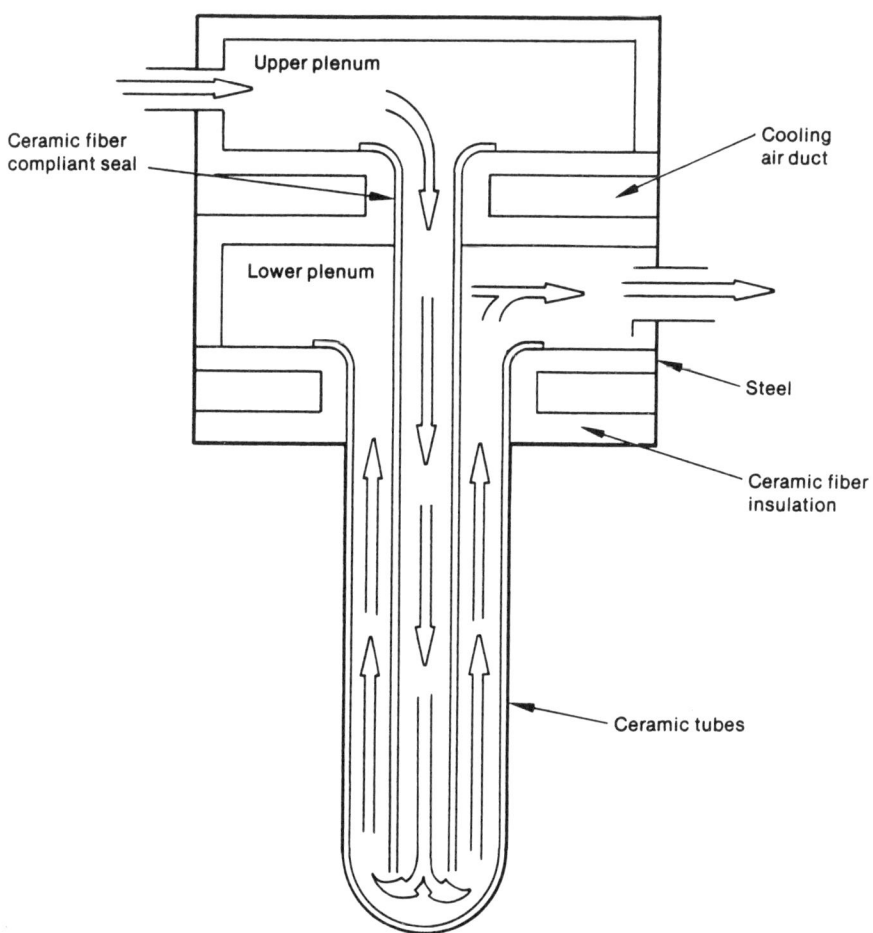

Figure 4. Babcock & Wilcox HTBDR.

Current Ceramic Heat Exchanger Designs 239

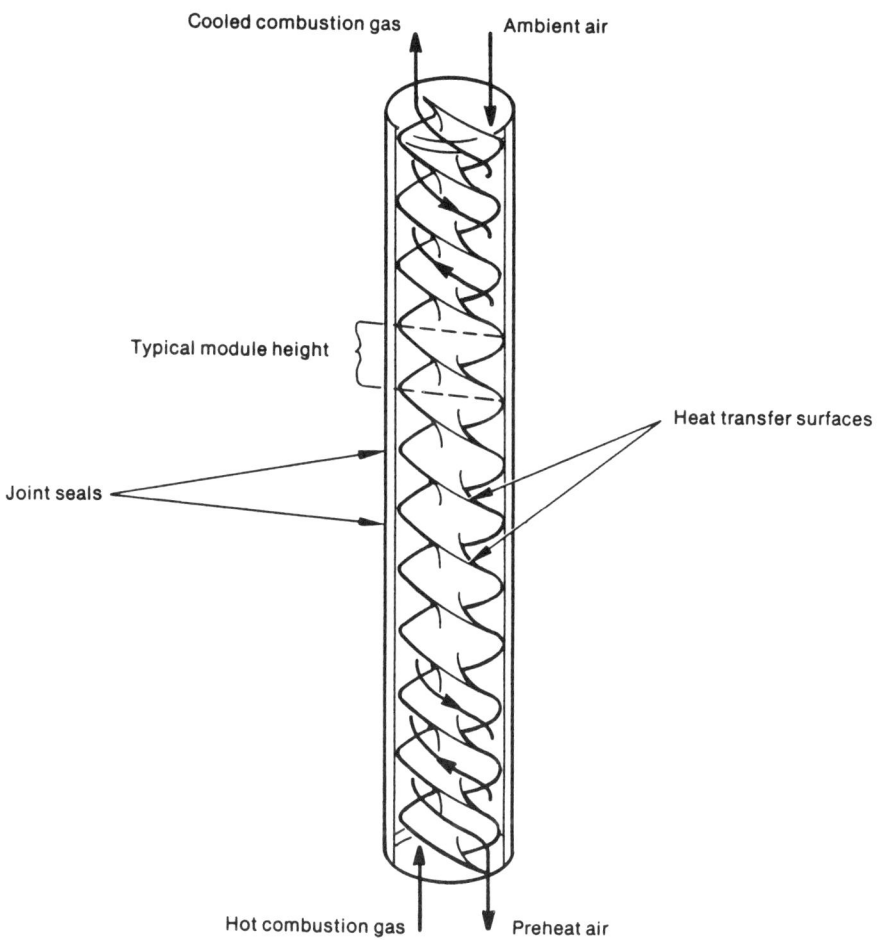

Figure 5. Schematic of the modular counterflow recuperator with a helical interface.

(ECP-3) and magnesia chromia spinel (X-81 and Unichrome) indicated potentially good material performance in a simulated soda-lime glass environment. Difficulties with this design are leakage between modules and fabrication problems (casting) with the double helix version and the finned, single-helix configuration. If these difficulties could be surmounted, long-term durability of the unit would still be unknown. More information can be found in Reference 13.

Solar Turbines International Design

Solar Turbines International has tested a tube-and-sheet recuperator designed to operate at a preheated air pressure of 100 psig vs. the usual 10 to 50 inches of water column (Figure 6). This heat exchanger is designed to be used for recuperating heat from coal-fired gas turbine exhaust streams. The tubes are manufactured from reaction-sintered SiC (NC 430) in 4- to 8-foot ligaments brazed together (with an alignment collar) to form 15- to 20-foot tubes. Airflow through the tubes is in a counterflow configuration and is ducted into and out of the exhaust stream by means of headers. The headers at the cold end of the tubes have a relaxing joint. A viscous glass seal at this joint tended to extrude from the joint at operating pressures. The next version used ball and socket joints to allow for thermal expansions and misalignments. Major difficulties with this design have stemmed from relatively small tolerances on sphericity and the need for hand-lapping each joint to minimize leakage. More information can be found in References 14 and 15.

Midland-Ross Design

The Midland-Ross/GRI segmented matrix heat wheel is designed for use with the relatively clean exhaust from gas turbines (Figure 7). This recuperator design uses segmented matrix elements extruded from MAS and cast in place with Thermosil 120. Interim mat is used as a resilient bonding media between the matrix assembly and the metal housing. This design also features axial labyrinth seals. Difficulties with this design are

Current Ceramic Heat Exchanger Designs

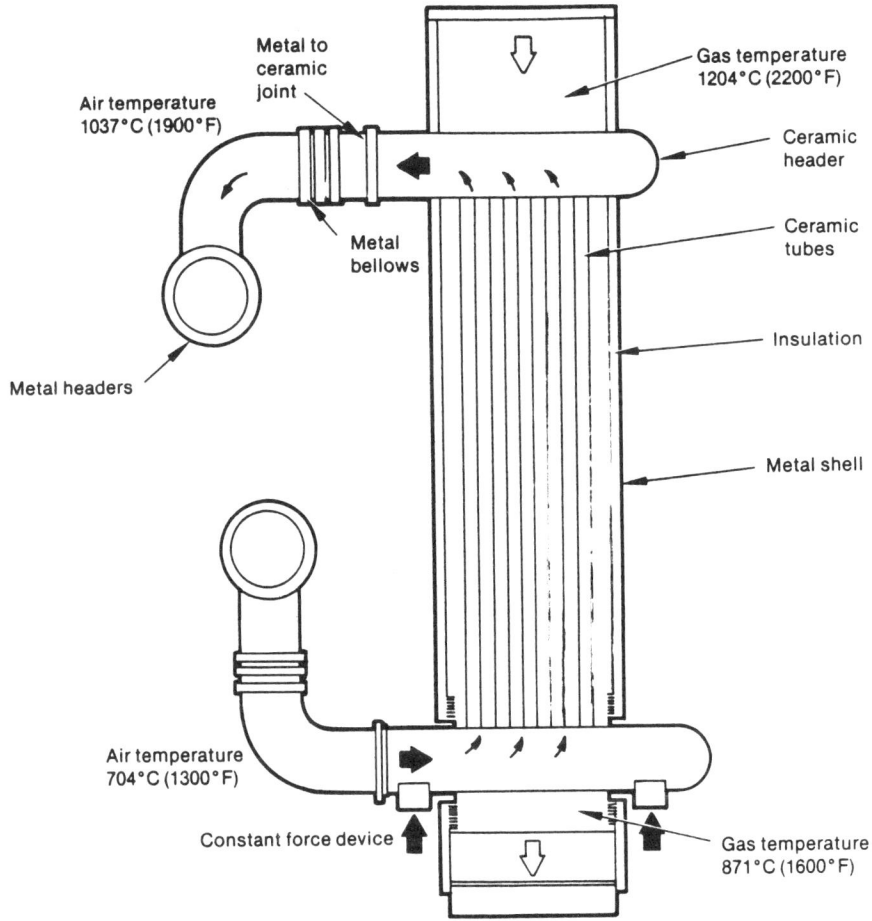

Figure 6. Cross-section of heat exchanger (Solar Turbines International).

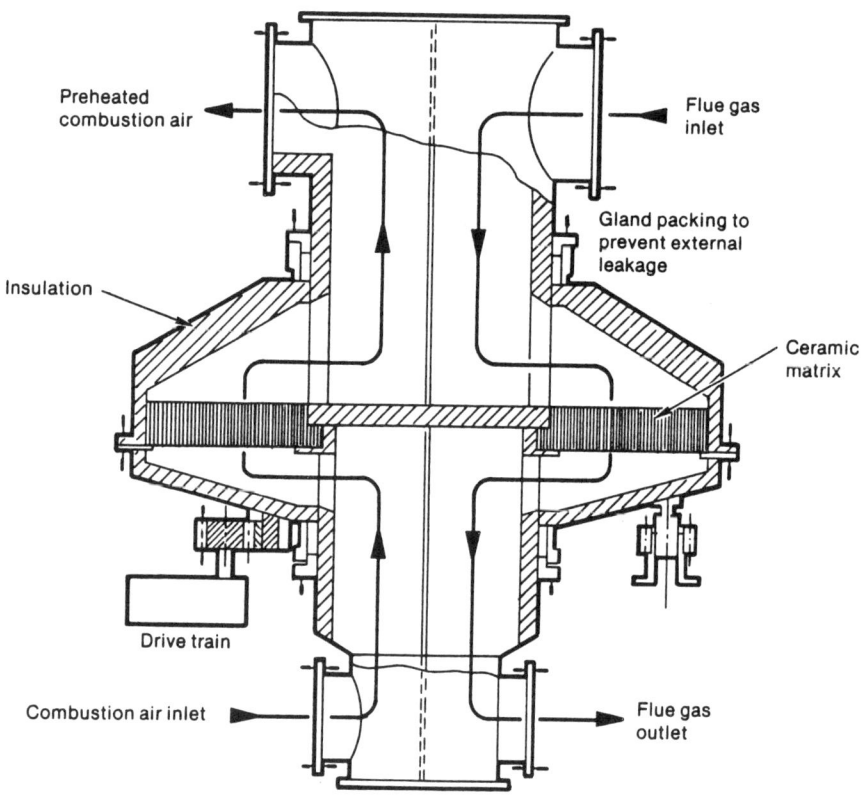

Figure 7. Schematic of the Heat Wheel.

seal leakage (labyrinth and stream-to-stream), possible problems with
matrix integrity, and severe fouling when used in dirty streams. More
information can be found in Reference 16.

Conclusions

The state-of-the-art and emerging high-temperature ceramic recuperator
designs offer significant improvements especially in terms of performance.
The more recent designs of Garrett and B&W offer particular improvements in
performance by the ability to produce 2000°F preheated combustion air.
Although the improved performance is related to their materials of construction, the material ability to perform adequately in the aluminum
remelt and glass melt furnace flue gases is unknown. None of the units
have been sufficiently tested to date in these waste streams to predict
long term durability, which is difficult to assess because of the variability in stream characteristics from different furnaces of the same
type. (See Reference 1 for stream characteristics.)

3. Assessment of Ceramic Recuperator Design Needs

It is clear from the previous section that there are many unknowns about the durability of currently available recuperators in the selected waste heat streams. By reviewing and assessing the design process for ceramic recuperators, as is done in the following section, it may be possible to determine what kind and how much information is needed by the designer to determine and/or assume durability.

<u>Design Process</u>

A knowledge of what a device is to do or how it is to perform is essential in order to design it. The more information that is known the better a design can be and the less iterations it will need to get final form. The design process can be normally broken down into a few broad substeps.

Assuming that a basic need has been identified, the first step is to establish performance requirements. When refined and approved, they become design requirements. This will be discussed further in this section. The second step is to identify and generate alternatives. This means to look at current industrial practice and state-of-the-art and identify appropriate configurations or generate new configurations that improve current industrial practice or advance the state-of-the-art. The third step is to develop a preliminary design. The details of this step will depend on the amount of material property data available. Generally, candidate materials are identified, a mathematical model is developed as a tool for preliminary sizing, analysis, and optimization; compatibility with system interfaces is assured; and preliminary cost estimates are generated. An economic analysis is the fourth step, the transition point at which the design gets the go-ahead or gets shelved.

Assuming the economic analysis is positive, prototype design, fabrication, and testing may be required as the fifth step, depending on the state

of the art. Preliminary materials testing may also be required if not much
data are available on the material at expected operating conditions.

The sixth step, prototype development, is a difficult step that is
necessary when exploring new areas of technology. Rarely does a first prototype work, but the information gathered on how and why failures occur
helps make the next iteration better. Sometimes material development is
required to achieve a good prototype. Slight material constituent ratio
changes may transform an unsuitable material into one eminently suitable
for the intended use. This is also the time for optimization of the
design, fabrication, and installation details. Perhaps a major redesign is
necessary and hence retesting is required. In any case, the design is
finalized in the last stages of this step.

Concurrent with prototype development, detailed structural and thermal
hydraulic analysis occurs, utilizing the mathematical models and algorithms
developed in step three and refined throughout the design. The final
analysis, of course, must wait for the final design to be approved and
should incorporate any recent changes and refinements.

The following sections address in more detail the two critical areas
of information vital to determining the direction and extent of the whole
design process: design requirements and available material data.

Design Requirements

In determining the design requirements of a high-temperature recuperator, it is necessary to consider what the user's (industry's) needs are
and what special requirements are imposed by the furnace to which the
recuperator will be attached. Since the main problem with current recuperators is their unknown durability in high temperature corrosive streams,
requirements related to durability will be focused on.

In the design and operation of recuperators, industry is mainly concerned with their performance, their durability, and their cost. The designer therefore needs to be concerned with everything from thermal performance and corrosion resistance to payback. In relation to durability, however, the designer needs to be concerned with corrosion resistance, mechanical strength, durability of the basic material and joints, and product reliability requirements.

The design requirements imposed by the furnace to which the recuperator is attached must include the ability to operate in the high-temperature, cyclic temperature, corrosive waste gas streams. Once a design configuration is selected with these requirements in mind, the necessary mechanical strength, corrosion resistance, and lifetime requirements for this design must be considered. These requirements will be affected by what types of steady-state and fluctuating conditions the recuperator is exposed to (such as startup, shutdown, upset, recovery, low fire, and high fire), especially how these conditions affect the flue gas chemistry. The structural design requirements or mechanical strength requirements are normally determined by the latter transient conditions which can cause severe thermal shock and expansion stresses. This is further complicated in the selected streams by material exposure to continuous high temperature corrosion. It follows that a material must be selected which satisfies these requirements. The material must satisfy the high temperature operating requirements by having high refractoriness, adequate high temperature strength, low thermal expansion coefficients with no sudden volume or phase changes and sufficient fracture toughness. It must also satisfy the requirements of ability to withstand long-term exposure to cyclic high temperature corrosive gases by having adequate thermal shock and thermal stress resistance, surface corrosion/oxidation resistance, and resistance to high temperature deformation mechanisms such as creep and slow crack growth. The lifetime reliability requirements of the candidate material once in service are very dependent upon these properties.

Thus far, designers have chosen conservative recuperator configurations which avoid any rigid mechanical strength requirements mainly by using compliant joints and thick materials. The effects of long-term stress corrosion (which are currently unknown), the effects of operation (especially thermal cycling with corrosion) on compliant joints (which

are currently unknown), and the desire for future flexibility in higher
stress designs may make these mechanical strength requirements more
stringent.

The other two aspects of durability which concern designers are the
ability to predict the desired lifetime and routinely assume a durability.
These requirements are indirectly affected by the waste gas streams, by how
the streams affect the ceramic materials so that the existing techniques
can be used or adapted to predict lifetime and assume a given durability
routinely.

Material Data Requirements

The material data requirements for a design can vary depending on the
approach selected by the designer, which may reflect the amount of material
data information available. Some combination of design approaches is
almost always applied mainly because of the lack of uniform or consistent
design data especially for new materials on unique applications.

The first approach used by designers is basically trial-and-error when
using materials for which little property data are available. The procedure is one of iterative fabrication and testing. Hague used this
approach as did GTE and Solar Turbines to some extent in designing the
units discussed in the preceding section.

A second approach applies the standard safety factor to the maximum
calculated stress. The material that has strength greater than the component stress by the factor of safety is selected. Due to the large variation or scatter in the strength of ceramic materials, the successful
application of the method to ceramic components requires larger factors of
safety than for metal components. Depending on whether the factor of
safety is large enough, some iteration may be necessary in fabrication and
testing. If the assumed factor of safety is much larger than it needs to
be, a conservative design will result. B&W and GTE used this approach.
Solar turbines also used this approach to a degree in designing the units
discussed earlier.

Currently, the most widely used design approach for structural ceramic materials is the probabilistic approach, where both flaw and stress distributions within the material are taken into account. This design approach is predicated on Weibull statistics. Thus, the performance of a given material is determined in a statistical manner which assumes that at a given load, failure will initiate at the most severe pre-existing flaw. The application of Weibull statistics is discussed further later in this report.

There are several advantages of the probabilistic design approach. First, the designer can apply this method to compute a probability of failure and compare this value to a predetermined or allowable quantity. Additionally, the material may be stressed closer to its actual limits (compared with the safety factor approach) without an excessive probability of failure. There are obviously some disadvantages of this method, one of which is poor definition of actual applied stresses and the need for a large statistical material data base; these deficiencies result in decreased component lifetimes. Garrett has used this design approach in the design of their HTBDR tube type recuperator described in the preceding section.

Perhaps the most theoretical design approach is the fracture mechanics approach. This method is predicated on the assumption that the structural material contains flaws which propagate when stressed to a specific level. The flaw or crack size, crack growth rate, and stress intensity at the flaw tip are required design parameters. Due to the statistical nature of strength in ceramic materials, a relatively large data base is required to apply this method. Other applications of linear elastic fracture mechanics to the design process are discussed later.

At the design stage, none of the currently used design approaches account for long term durability of the recuperator. The long-term exposure and stress data approach would require data for different stresses, temperatures, and environments over time without any assumptions concerning the mode of failure. Thus, this design approach would allow for multiple failure modes. For a given environment, a statistically significant number of data points would be needed for a range of stresses, temperatures, and times such that the resultant data base could be as large as the probabilistic data base or the linear elastic fracture mechanics data base. The problem with this approach is that the data are only good for the environment tested. The benefits of this approach are that the designer can be more confident about the long-term high-temperature behavior of the chosen material so that he/she can design less conservatively.

Conclusions

Given the requirements of industry to have a durable recuperator and the cyclic high-temperature corrosive nature of the waste gas streams, the designer needs information on the long term durability of the recuperator's materials of construction, especially if higher-stress designs are to be developed in the future.

The designer may use any number of approaches to obtain durability information, from iterative fabrication and testing to the use of long term exposure and stress test data:

- o Trial and error fabrication and testing, with little materials property data;
- o Application of standard safety factor to the maximum calculated stress;
- o Probabilistic treatment of data, considering flaw and stress distributions;
- o Use of fracture mechanics data; and
- o Use of long-term exposure and stress test data.

The iterative, or trial and error, approach could involve much time and expense with little information that could be extrapolated to the design of another ceramic recuperator. The long-term exposure and stress test data approach could save money in design, in the long term, and be applicable to more than one ceramic recuperator design, but the results may be applicable to only the test environment. With this approach, the designer may predict a given lifetime, and provided that sufficient correlations are made, a technique for routinely assuming this durability may be found. Other approaches using non-environmental short-term test data would provide information that may be extrapolated, within a margin of error, to other ceramic designs, but some extra-cost iteration may be necessary depending on long-term operational durability results.

4. Ceramic Materials for High Temperature Heat Exchanger Application

In relation to durability, the designer is concerned with corrosion resistance and the time dependence of mechanical strength of the ceramic material so that a lifetime may be predicted and a product reliability assumed. The amount of information needed by the designer was discussed in the preceding section. Candidate available ceramics and the amount and quality of design information available for these materials are discussed in this section. Some possible methods for improving the durability of candidate ceramics are also discussed.

<u>Selection of Candidate Materials</u>

The only known materials having adequate physical and mechanical properties, oxidation/corrosion resistance, and fabricability for ceramic recuperator tubes are silicon carbide, silicon nitride, and oxide ceramics. Candidate materials for specific industrial furnaces are shown in Table 3. Silicon carbide is the prime candidate for steel soaking pits and aluminum remelting furnaces. The properties that recommend silicon carbide for use in these environments are mechanical strength, oxidation/corrosion resistance, and resistance to thermal shock, the latter being particularly important in cases where thermal cycling frequency is high. The corrosion behavior of silicon carbide in soaking pit and aluminum remelt furnace environments is not well known. The fact that alkali oxides and sulfates are present in the flue gases of steel soaking pits and aluminum remelt furnaces is cause for concern because silicon carbide corrosion (to be described) was partly attributed to alkalies in a glass furnace environment. Although some authorities doubt the suitability of this material for this application, recent corrosion tests in soaking pit and aluminum remelt furnace flue gases have been encouraging as will be described later. Silicon nitride is also a candidate material for these furnaces unless oxidation resistance proves to be inadequate.

Silicon carbide is not recommended for use in glass melting furnaces because of poor corrosion resistance to this environment. The typical flue gas temperature in a glass furnace is 1600°C, and the gases contain both

TABLE 3. CANDIDATE CERAMIC MATERIALS FOR RECUPERATORS FOR INDUSTRIAL FURNACES

Furnace	Material
Steel soaking pit	Silicon carbide[a]
	Silicon nitride
Aluminum remelting	Silicon carbide[a]
	Silicon nitride
Glass remelting	Alumina
	Mullite
	Stabilized zirconia
	Zirconium silicate

a. Types include sintered-α, sintered β, recrystallized, reaction sintered (siliconized), nitride-bonded, oxide-bonded, and oxynitride-bonded.

particulates and vaporized species of the glass batch as well as fuel impurities. Fouling and corrosion of ceramic recuperators are potentially serious problems. High-purity alumina and zirconia-based ceramics are candidate materials for glass furnace waste heat streams.

Alumina (Al_2O_3) is very corrosion resistant and wear resistant; however, this property is dependent on its purity. The presence of small amounts of certain impurities in an alumina ceramic affects the desirable high temperature mechanical and chemical properties. SiO_2 especially will cause an alumina ceramic to creep at elevated temperatures. Although its thermal conductivity is lower than SiC, its chemical stability in a corrosive waste stream is better. The primary problem with Al_2O_3 is poor thermal shock resistance; therefore, an alumina recuperator cannot be thermally cycled.

Stabilized zirconia,[17] in comparison with alumina, is a more refractory material, has a lower flexural strength, thermal conductivity, and modulus of elasticity, and has a higher coefficient of expansion. The particularly low modulus of elasticity relative to that of alumina results in good thermal shock resistance. However, stabilized zirconia has not yet been evaluated by corrosion testing as a potential candidate for high temperature heat recuperation. Partially stabilized zirconia, while exhibiting higher fracture toughness than fully stabilized zirconia, is not a candidate material because of degradation of properties at temperatures above approximately 1000°C.

Description of Candidate Silicon Carbides

A brief description of six types of silicon carbide ceramics is presented here to aid in understanding the greatly different properties of silicon carbides.

Sintered-α and Sintered-β SiC. These materials are prepared by sintering highly-pure powders of SiC to which small amounts (<1 wt%) of sintering aids have been added. The powders are consolidated by cold pressing, slip casting, extrusion, etc. into the desired shape, then sintered to high density (typically >98% of theoretical density). The materials are stronger than other SiC ceramics and impermeable to gases because of the relatively low porosity and because only small amounts of second phases are present. In sintered-α, the dominant crystalline phase is alpha (hexagonal), which is the high temperature form of SiC. Beta phase (cubic) transforms to alpha during extended heating at high temperatures (~2000°C).

Recrystallized SiC. This material is also highly pure SiC (>99 wt% SiC). Powders are consolidated into desired shapes without addition of sintering aids, then pressureless sintered. Sintering results in densification by interparticle diffusion and grain growth. The sintered body, however, contains as much as 20 vol% residual porosity and, therefore, can be permeable to gases. The strength is substantially less than that of sintered-α or sintered-β.

Reaction Sintered or Siliconized SiC. This material is fabricated by sintering a mixture of primary SiC grains with additions of silicon and carbon. The silicon and carbon react during sintering to form secondary SiC grains. Alternatively, silicon is infiltrated into a presintered SiC and carbon body; a subsequent sintering causes silicon to react with carbon to form secondary SiC grains. These materials are characterized by primary SiC grains in a matrix of smaller secondary SiC grains, free silicon, and porosity. The amount of free silicon and porosity can be minimized by fabrication procedures.

Nitride-Bonded SiC. A mixture of primary SiC grains and smaller silicon grains is consolidated, then heated in a nitrogen atmosphere to convert silicon to Si_3N_4. Self-bonding among Si_3N_4 grains and between Si_3N_4 and SiC grains is responsible for the material's strength. The structure consists of primary SiC grains in a Si_3N_4 matrix containing as much as 15 vol% porosity. The material is generally considerably weaker than previously mentioned SiC ceramics, and is permeable to gases.

Oxide-Bonded SiC. This material is fabricated by sintering a mixture of primary SiC grains and a silica-rich oxide phase. Sintering causes bonding among the silica grains and between the silica and primary SiC grains. The structure consists of primary SiC grains in a highly porous (15 to 18 vol% porosity) oxide matrix. This material is generally the weakest of all SiC ceramics and is permeable to gases.

Chemical Vapor Deposited SiC. This material is generally deposited as a coating on a suitable substrate; however, free-standing bodies can also be prepared. Various gaseous precursors are used in deposition. Two commonly used gaseous mixtures are CH_3SiCl_3-H_2 and $SiCl_4$-H_2-CH_x. Methyltrichlorosilane (CH_3SiCl_3), which contains both Si and C, is thermally decomposed in an H_2 atmosphere. Silicon tetrachloride ($SiCl_4$) is reduced by H_2 in the presence of a hydrocarbon such as methane. These reactions are usually conducted at temperatures above 1400°C. The deposits typically have a columnar grain structure with long axes perpendicular to the substrate. Under optimum deposition conditions, this material is highly dense and pure.

Available Mechanical Properties

The following paragraphs assess the quantity and quality of mechanical property information available to the designer. Properties available from the manufacturers' data and from the literature are reviewed. Additional available property information on ceramics exposed to corrosive environments is also discussed.

Properties Available from Manufacturers. Ideally, a large and diverse compilation of properties data should be available for use in recuperator design. Unfortunately, information related to lifetime prediction is incomplete and inconsistent for high temperature ceramic recuperator materials. Manufacturers typically provide properties measured at ambient temperature, supplemented to varying degrees with higher temperature data. The designer's task is further complicated by the fact that the ceramics which are usable in recuperator applications cannot be placed in generic groups having uniform properties. To varying degrees each manufacturer's product is unique with properties dependent on the feed material, a particular manufacturing process, and surface finishing operations, among others. Also, chemical interaction with operating environment can alter the properties. The designer must, therefore, use the available properties data with discretion.

Properties information supplied by manufacturers for as-fabricated ceramic materials is included in Appendix A and summarized in Table 4. The term as-fabricated, as used here, refers to material that has not been altered by a thermal treatment or exposure to a corrosive environment; however, the definition does not exclude testing in air at temperatures above ambient. The data in these tables reflect the readily accessible manufacturers' information for state-of-the-art ceramic materials. Data are quite limited for some materials. Representatives of two major manufacturers emphasize that some materials are still improving as changes are made in feedstocks and production methods.

TABLE 4. SUMMARY OF MECHANICAL PROPERTIES OF CERAMIC MATERIALS PROVIDED BY MANUFACTURERS

Material	Designation	Manufacturer	Flexure strength MPa	Weibull modulus	Tensile strength MPa	Fracture toughness MPa·m^(1/2)	Compressive strength MPa	Young's modulus GPa	Bulk modulus GPa	Shear modulus GPa	Poisson ratio
Sintered SiC	Sintered-α	The Carborundum Co.	459	12.3		4.6		406		178	0.142
Reaction sintered SiC	Hexalloy KT	The Carborundum Co.	383	10		4.97		332		147	0.127
Reaction sintered SiC	SC-2	Coors Porcelain Co.	517		307		2480	395			0.19
Siliconized SiC	NC-430	Norton Co.	310					400			
Siliconized SiC	Crystar - HD	Norton Co.	300			3.5					
Siliconized SiC	Crystar- K	Norton Co.									
Recrystallized SiC	Crystar	Norton Co.	46					280			
Nitride-bonded SiC	Cerasurf TN-15	Coors Porcelain co.	59					210			
Nitride-bonded SiC	Cryston CN 178	Norton Co.	100								
Nitride-bonded SiC	Cryston CN 983	Norton Co.	24								
Oxynitride-bonded SiC	Crystolon CN 163	Norton Co.	29								
Oxide-bonded SiC	Crystolon CN 181	Norton Co.									
Aluminum oxide	AD-999	Coors Porcelain Co.	517		310		3792	386	228	158	0.22
Aluminum oxide	AD-995	Coors Porcelain Co.	379		262		2260	372	228	152	0.22
Mullite	Mullite	Coors Porcelain Co.	186				551	155			
Cordierite	CD-1	Coors Porcelain Co.			124			138			0.316

a. At 25°C.

Although all mechanical property data are potentially useful, certain values are critical for design purposes. Among these are flexure and/or tensile strength and associated Weibull modulus, Young's modulus, stress rupture, creep strength, and fracture toughness. The limited availability of data on these properties is apparent on examination of the tables. For most materials, values at ambient temperature for all of these properties are not available, and only in a few cases are elevated temperature values available. Table 4 summarizes mechanical properties (at 25°C) supplied by manufacturers. The table shows that flexure strength data are available for most materials; otherwise, the table is mostly incomplete. The Weibull modulus (required for conservative use of strength values) and fracture toughness (used for lifetime prediction) are mostly not available. More high temperature data exist for sintered-α and Hexoloy KT SiC than for other materials (Tables A-1 and A-2). Most of the data in the tables are presented in SI units; however, an abbreviated list of factors is presented in Appendix B in Table B-1 for convenience in converting to other units.

The data for as-fabricated materials in the tables should be used cautiously. The manufacturers obtained this data with samples specially prepared for testing. For example, test samples were cut from larger pieces, surfaces were ground, edges were beveled, etc. Strength values determined with such samples might not be attained with recuperator components such as tubes, which are used without the benefit of machining operations that might significantly decrease the population and size of flaws on as-fabricated surfaces. The manufacturers' strength data can be used to rank materials for a particular application, but the actual performance of the materials might indicate a different ranking order.

Properties Available from the Literature. Since the mechanical property data supplied by manufacturers is limited, the technical literature has been examined for additional information. These data supplement and extend the data provided by manufacturers; however, the relevancy of the data to currently fabricated and marketed materials must be determined.

Flexure testing and Weibull statistical analysis are commonly used to characterize the strength of ceramic materials. Weibull statistical analysis, which is discussed later in this report, suggests that a large number of tests are required for results to be statistically significant. This is especially true for flexure test results because the volume of material exposed to fracture stress is small relative to the specimen size. Flexure test results for sintered-α and siliconized SiC reported by Larsen and Adams[18] and by Coombs, Kotchick, and Weidhaas[19] are presented in Tables 5, 6, and 7. These results were obtained using small bend bars (for example, 50 x 6 x 3 mm tested in 3- or 4-point bending). Values are given for temperatures up to 1500°C for sintered-α and up to 1350°C for siliconized SiC. The mean flexure strengths reported by these investigators are generally lower than those reported by the manufacturers (Tables A-1, A-2, and A-3). The reason for the discrepancy is not apparent for the data of Larsen and Adams (Reference 18). Possible reasons are differences in materials, specimen preparation, and test methods. The mean strength values reported by Coombs et al.[19] are probably lower than manufacturers' values because the tension surface of the former's specimens were not ground. Most likely flaws in the as-fabricated surfaces of the specimens precipitated failure at the stresses shown.

Flexural strengths have also been obtained using ring specimens,[20] which provides the benefits of using test materials fabricated by the same methods and having the same geometry as heat exchanger tubes. Ring fracture strengths were obtained for sintered-α and siliconized SiC at temperatures to 1260° and 1371°C, respectively. The mean strengths, presented in Tables 8, 9, and 10, were substantially lower than flexure strengths provided by manufacturers (see Appendix A). Since tension surfaces of the specimens were not ground, the low values probably reflect the effects of flaws in as-fabricated surfaces.

Another method of characterizing ceramic materials is stress rupture testing (or static fatigue). This test determines the time to failure at constant temperature of a group of specimens stressed at various levels below the stress that causes instantaneous fracture. Fracture occurs by

TABLE 5. FLEXURE TEST RESULTS FOR SINTERED- SILICON CARBIDE[a,b]

IDENTIFICATION

Manufacturer -- The Carborundum Company
Designation -- Sintered-α (1981)
Nominal Composition -- 99 wt % SiC

Test Temperature °C	Number of Specimens	Mean Density[c] Mg/m^3	Surface Finish μ in.	Mean Strength MPa	Weibull Modulus	Mean Strain to Failure in./in.×10^{-3}	Young's Modulus GPa
25	9	3.117	57	325.7	6.0	0.84	386
1000	9	3.100	57	392.3	8.1	1.12	352
1250	9	3.105	67	407.1	13.4	1.43	281
1500	9	3.111	56	311.4	6.5	1.07	290

a. Source: Reference 18.
b. 4-point bending in air.
c. Theoretical density = 3.217 Mg/m^3.

TABLE 6. FLEXURE TEST RESULTS FOR SILICONIZED SILICON CARBIDE[a,b]

IDENTIFICATION

Manufacturer	-- Coors Porcelain Company
Designation	-- Reaction Sintered SC-2 (1981)
Nominal Composition	-- Not given

Test Temperature °C	Number of Specimens	Mean Density[c] Mg/m^3	Surface Finish μ in.	Mean Strength MPa	Weibull Modulus[d]	Mean Strain to Failure in./in.×10^{-3}
25	5	3.097	28	313.1	--	0.83
1000	4	3.085	26	389.2	--	1.00
1200	4	3.089	27	377.7	--	1.31
1350	4	3.089	26	303.8	--	1.32

a. Source: Reference 18.

b. 4-point bending in air.

c. Theoretical density = 3.099 Mg/m^3.

d. Not given.

slow crack growth, and examination of the fracture surface often reveals flaws that cause crack formation. Static fatigue of SiC has been reviewed by Quinn.[21] Various investigators have found that SiC is susceptible to static fatigue; however, its resistance is considered good since failure only occurs at stresses near the fast-fracture strength.

The static fatigue behavior of sintered-α SiC fabricated during 1978 and 1980 is shown in Figure 8. Within the scatter of the time dependence of these data, Quinn and coworkers determined that stresses must exceed 70% of fast fracture strength to cause static fatigue failure. The conclusion from a larger body of test data obtained at high temperatures was that stresses must exceed 65% of average room temperature strength to cause static fatigue failure. Two characteristics of the data in Figure 8 should be noted. An indication of the limiting stress levels to yield a lifetime of

TABLE 7. FLEXURE TEST RESULTS FOR SILICONIZED SILICON CARBIDE IN AIR[a]

IDENTIFICATION

Manufacturer -- Norton Company
Designation -- NC-430 (1980 and 1981)
Nominal Composition -- Not given

Test Temperature °C	Number of Specimens[b]	Surface Condition[c]	Displacement Rate in./min.	Mean Strength MPa	Weibull Modulus
21	15	G	0.02	199[d]	19.1
21	19	AP	0.02	146[d]	10.3
899	5	AP	0.02	161[d]	9.5
899	7	AP	0.002	187[d]	5.1
899	10	AP	0.0002	192[d]	14.3
1060	5	AP	0.02	150[d]	8.5
1060	4	AP	0.0002	185[d]	7.9
1260	30	AP	0.02	187[e,f]	9.4
1260	30	AP	0.0002	209[e]	12.0
1343	10	AP	0.02	161[d]	7.6
1343	7	AP	0.002	181[d]	30.1
1343	10	AP	0.0002	190[d]	14.0
1343	8	APO	0.0002	179[d]	8.4

a. Source: Reference 19.

b. Size: 0.25 in. x 0.125 in. x 2 in.

c. G -- tension surface ground to 320 grit.
   AP -- Tension Surface as-processed.
   APO -- specimens oxidized in air for 250 h at 1260°C.

d. 3-point bend test normalized to 4-point bend test area (1980 material).

e. 4-point bend test with 1.5 on. outer span, 0.75 in. inner span (1981 material).

f. Oxidized for 1 h at 1260°C before testing.

TABLE 8. RING FRACTURE TEST RESULTS FOR SINTERED-α SILICON CARBIDE[a]

IDENTIFICATION

Manufacturer -- The Carborundum Company
Designation -- Sintered-α (1980 or earlier)
Nominal Composition -- Not given

Test Temperature °C	Number of Specimens	Mean Strength MPa	Weibull Modulus
25	32	200	6.0
815	4	154	27.8
926	5	198	6.1
1037	5	198	5.7
1260	5	257	5.7

a. Source: Reference 20.

TABLE 9. RING FRACTURE TEST RESULTS FOR SILICONIZED SILICON CARBIDE[a]

IDENTIFICATION

Manufacturer -- The Norton Company
Designation -- NC-430 (1980 or earlier)
Nominal Composition -- Not given

Test Temperature °C	Number of Specimens	Mean Strength MPa	Weibull Modulus
25	30	182	9.3
816	8	157	6.5
926	4	216	8.9
1036	4	183	8.9
1260	4	181	8.3
1371	4	113	9.1

a. Source: Reference 20.

TABLE 10. RING FRACTURE TEST RESULTS FOR SILICONIZED SILICON CARBIDE[a]

IDENTIFICATION

Manufacturer -- The Carborundum Company
Designation -- KT and Super KT (1980 or earlier)
Nominal Composition -- Not given

Test Temperature °C	Number of Specimens	Mean Strength MPa	Weibull Modulus
		KT	
25	15	144	5.6
1204	14	125	11.1
		Super KT	
816	8	269	8.7
926	4	228	8.2
1037	4	290	11.9
1260	4	250	11.9
1371	4	107	7.2

a. Source: Reference 20.

1000 h was determined for 1978 and 1980 vintage sintered-α SiC. Little data exist for tests of this duration. In general, the 1980 material was stronger, but the data for 1978 material showed less scatter.

The results of tests to observe the effect of temperature on stress rupture are shown in Figure 9. Samples were tested at 1350, 1400, and 1500°C in air. The trend of the stress-time relationship up to the 100 h maximum time of these tests was very similar to that found at 1200°C (see Appendix A). A strong dependence on temperature was not revealed.

The results of stress rupture tests at 1200°C in air for NC-433 and NC-435, which are siliconized silicon carbides, are shown in Figure 10. Also shown are data for Silcomp CC, which is fabricated by infiltrating a precursor plate of carbon fibers with silicon, then heating to form silicon

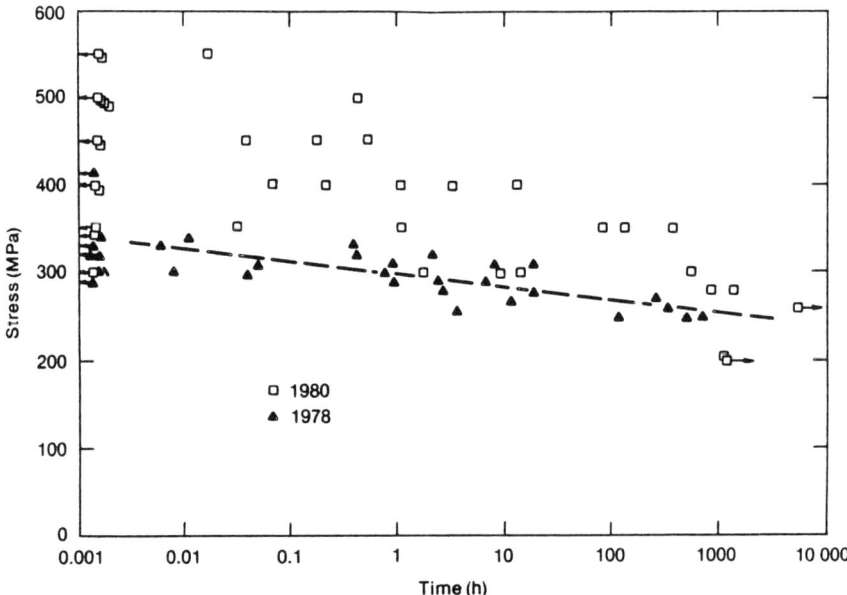

Figure 8. Flexural stress rupture of 1978 and 1980 vintages of sintered alpha silicon carbide at 1200°C in air. The line is for the 1978 data. Taken from Reference 19.

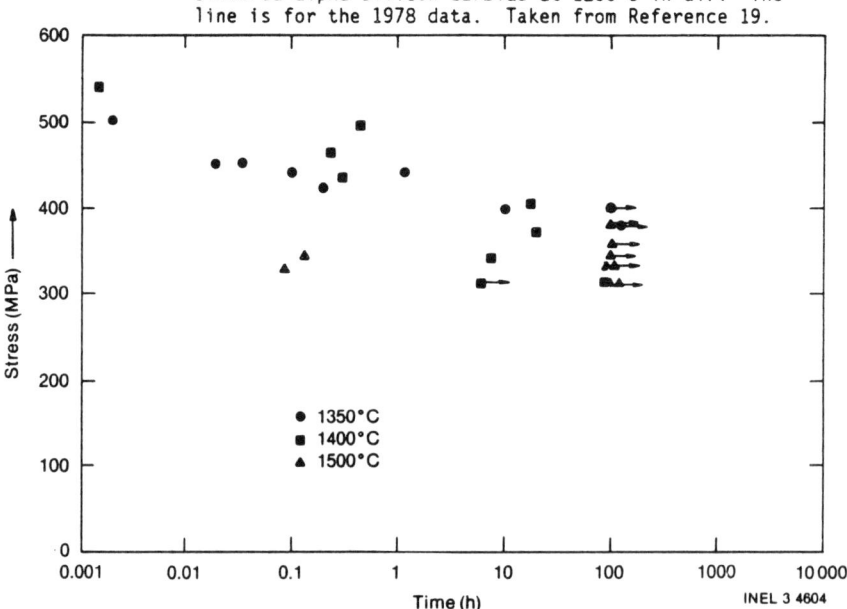

Figure 9. Flexural stress rupture of sintered alpha silicon carbide in air. Taken from Reference 19.

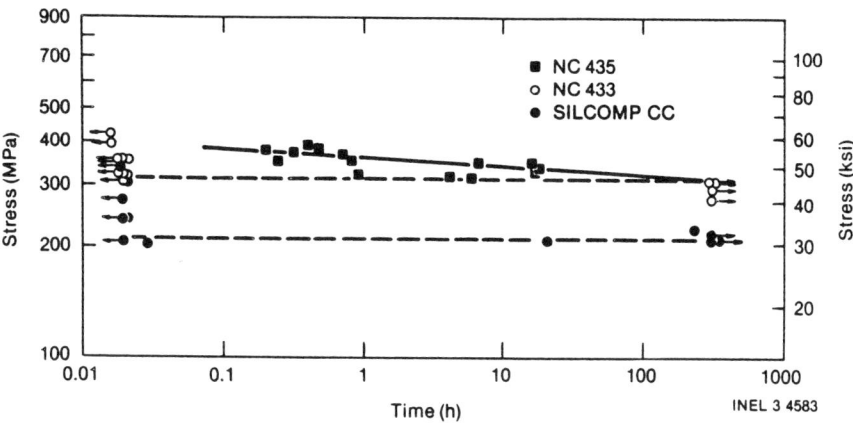

Figure 10. Flexural stress rupture of NC-435, NC-433, and Silcomp CC, siliconized silicon carbides at 1200°C in air. Taken from Reference 21.

carbide by reaction of elemental carbon and silicon. The data for siliconized SiC appear to have the same form as those shown previously in Figure 9. Unfortunately, the test duration was only several hundred hours.

Coombs, Kotchick, and Weidhass[19] also reported on the stress rupture properties of NC-430 siliconized SiC. Stress rupture and permanent midspan deflection results are given in Table 11. Material fabricated during 1980 and 1981 were included in these tests. The results indicate that NC-430 does not undergo slow crack growth at 1260°C in air. A greater than 90% survival rate was observed for samples stressed to 70% of a baseline value of 193 MPa (28.0 ksi) established in earlier tests on 1980 material. In general, the NC-430 prepared during 1980 performed better than that fabricated in 1981.

Cycling of load or temperature might have an appreciable effect on stress rupture behavior, but little data exists. Coombs, Kotchick, and Weidhaas (Reference 19) tested at 1260°C the effect on NC-430 of a cyclical

TABLE 11. STRESS RUPTURE RESULTS FOR SILICONIZED SILICON CARBIDE TESTED AT 1260°C IN AIR

IDENTIFICATION

Manufacturer -- Norton Company
Designation -- NC-430 (1980 and 1981)
Nominal Composition -- Not given

Number of[b] Specimens	Stress Level MPa / %[c]		Number of Specimens that Broke in the Indicated Time				Number of Specimens that Did Not Break in 500 h	Average Midspan Deflection for Unbroken Specimens	
			0 min	0.5 min/ 3 min	242.5 h	388 h		in.	mm
			1980 material						
5	174-188	90-97.5	4				1	0.0075	0.19
14	154-170	80-88	5				9	0.0960	2.44
6	135-150	70-78					6	0.0625	1.59
			1981 material						
14	174	90		11		1[d]			
10	154	80		5	1				
10	135	70		2	4	1[e] 2[e]	6	0.0026	0.07

a. Source: Reference 19.
b. Size: 0.25 in. x 0.125 in. x 2 in.; as-processed surface in tension.
c. Percent of mean flexure strength.
d. Failure caused by sonic shock wave created by adjacent specimen's failure on initial loading.
e. Failure due to furnace controller malfunction.

load varying from 0.1 to 1.0 times 70% of the previously established average fracture strength of 193 MPa (28.0 ksi). Each sample was tested for 1 h at a loading and unloading frequency of 1 Hz during which none of the samples failed. Subsequently, each sample was broken in flexure. The results of flexure testing, shown in Table 12, reveal an average strength of 214 MPa (31.0 ksi), which was an increase over the initial average strength. Additional testing will be required to establish cyclical fatigue limits as a function of time, frequency, temperature, and stress.

Larson and Adams (Reference 18) measured the flexural creep behavior in air of sintered-α (Carborundum 1981) and siliconized SiC (Coors SC-2) at 1500 and 1350°C, respectively. The results at incremental stress levels appear in Tables 13 and 14. Very low creep rates were found for sintered samples. The SC-2 exhibited somewhat higher creep rates at a given stress level even though tested at a lower temperature. The difference in behavior is probably due to the continuous silicon metal phase in SC-2.

In summary, an examination of the quantity and quality of mechanical property data available from manufacturers and other sources reveals that the information needed for conservative design is not available. Tensile and/or flexural strength with Weibull modulus at elevated temperatures are minimum requirements, but these are not available for most materials. In addition, stress rupture properties, creep strength, and fracture toughness at elevated temperatures would allow more efficient design, that is, without the use of excessively thick sections to compensate for low reliability. It should be noted that all of the referenced data were obtained in a fairly benign atmosphere (air). Corrosive environments can produce different results as described in the following paragraphs.

<u>Properties of Materials Exposed to Corrosive Environments</u>. In most applications ceramic recuperator materials will be exposed to corrosive species derived from fuels, furnace refractories, and products being processed. The high temperature mechanical properties of materials might degrade significantly as a result of corrosion by these species. Obviously, materials of high strength are desirable in cases where strength degradation occurs.

TABLE 12. CYCLIC FATIGUE AND FLEXURE TEST RESULTS FOR SILICONIZED SILICON CARBIDE TESTED AT 1260°C IN AIR[a]

IDENTIFICATION

Manufacturer -- Norton Company
Designation -- NC-430 (1980)
Nominal Composition -- Not given

Total Cycles[b]	Flexure Strength[c] MPa
3600	219
3600	228
3600	192
3600	214
3600	224
3600	208
3600	190
3600	216
3600	214
3600	232
	(214 mean value)

a. Source: Reference 19.

b. Minimum and maximum stresses were 13 and 31 MPa applied at a frequency of 1 Hz; no failures occurred.

c. 4-point flexure test after cyclic loading test.

The following paragraphs discuss the effects of several waste heat streams or materials. It should be noted, however, that these tests do not place the material under stress during exposure, so they do not fully duplicate actual operating conditions.

Steel Soaking Pit Exposure--Siliconized SiC was selected as the material for B&W's primary heat exchanger tubes. The reference material is NC-430 and alternates are KT and sintered-α SiC. These materials, along with high-purity alumina and mullite, were subjected to a corrosion test in

TABLE 13. CREEP TEST RESULTS FOR SINTERED-α SILICON CARBIDE AT 1500°C[a]

IDENTIFICATION	
Manufacturer	-- The Carborundum Company
Designation	-- Sintered-α (1981)
Nominal Composition	-- >99 wt % SiC

Applied Stress MPa	Steady State Creep Rate $h^{-1} \times 10^{-6}$	Duration at Each Stress h
70	5.75	24
103	2.07	24
138	5.83	24
138	6.58	24
172	~8.6	15.3[b]
172	--	0[b]
207	--	1.5[b]
241	--	0[b]

a. Source: Reference 18.
b. Sample failed.

a steel soaking pit for 36 days at a maximum temperature of 1290°C.[22] During the test, the pit operated normally including thermal cycling and use of hot topping compounds on billets being heated. At the conclusion of the test, the materials, in the form of tubes, were examined ceramographically and measured to determine changes in C-ring fracture strength and coefficient of thermal expansion.

After the test, the materials had a glassy slag surface coating, which consisted of a silicate containing Al, Ca, Fe, and K. Dimensional measurements and ceramographic examination revealed no significant corrosion in the form of surface recession or microstructural alterations. The mullite tube, however, was severely cracked. Neither the mullite nor alumina tubes

TABLE 14. CREEP TEST RESULTS FOR SILICONIZED SILICON CARBIDE AT 1350°C[a]

IDENTIFICATION	
Manufacturer	-- Coors Porcelain Company
Designation	-- Reaction Sintered SC-2
Nominal Composition	-- Not given

Applied Stress MPa	Steady State Duration at Creep Rate $h^{-1} \times 10^{-6}$	Each Stress h
35	~0.3	24
70	2.93	24
70	2.58	24
103	6.17	24
117	4.08	24
138	8.75	24
207	24.9	24
207	12.8	24

a. Source: Reference 18.

were subjected to additional measurements. The C-ring fracture strengths of the silicon carbide ceramics at 1100 and 1300°C revealed that no significant strength degradation occurred in most cases (data shown in Table 15). The strengths of exposed materials were almost as high or higher in some cases, sintered-α tested at 1300°C being an outstanding exception. Increases in strength were attributed to blunting of surface flaws by oxidation or by the glassy slag that deposited on the surface.

The results of this test, although encouraging, can only be construed as indicative of materials behavior for short term exposure. A much longer exposure time is needed to reveal the effects of corrosion on mechanical properties. A new test to be conducted for 6 to 12 months is presently being implemented.

TABLE 15. C-RING FRACTURE STRENGTHS OF SILICON CARBIDE EXPOSED TO STEEL SOAKING PIT FLUE GASES

Type of Material	Designation	Manufacturer	Test Temperature °C	C-Ring Fracture Strength, MPa[c]		Change, %
				As-Received	Exposed	
Reaction sintered SiC	CS-101	Norton Company	1100	103 (5)	119 (5)	15.5
			1300	85 (4)	117 (4)	36.0
Reaction sintered SiC	KT	Carborundum Company	1100	166 (5)	153 (4)	-7.8
			1300	152 (5)	139 (4)	-8.6
Siliconized SiC	NC-430	Norton Company	1100	152 (5)	152 (5)	0
			1300	168 (4)	143 (4)	-14.9
Sintered SiC	Sintered-α	Carborundum Company	1100	257 (4)	284 (2)	10.5
			1300	252 (5)	213 (2)	-15.5

a. Source: Reference 22.
b. 36 days of normal furnace operation; maximum temperature 1290°C.
c. Average strength for the number of specimens shown in parentheses.

Aluminum Remelt Exposure--Tubular specimens of recrystallized, siliconized, reaction sintered, and nitride-bonded silicon carbide were installed in an aluminum remelt furnace (fueled by natural gas with $NaCl-KCl-Na_3AlF_6$ and Cl_2 flux) for 1585 hours total, 565 hours of which was normal furnace operation at a maximum temperature of 1250°C. The materials developed a glassy slag surface coating,[23] which was molten during the exposure.

The slag contained major concentrations of K, Na, and Si and large amounts of Ba, Ca, Cu, Fe, and Zn. Ceramographic examination revealed no significant corrosion at the interface between the various materials and the slag layer as a result of this relatively short exposure (565 h). The nitride-bonded SiC (C/75C) was severely cracked as a result of the exposure. Possibly, the K and Na in the glassy slag would eventually attack the free silicon phase in siliconized and reaction sintered SiC materials, but no evidence of such a reaction was observed after this short exposure.

An exposure test of metallic alloys and SiC ceramics was conducted in the flue gases of another aluminum remelt furnace to identify suitable materials for recuperator applications. The flue gases contained chloride and fluoride species typical of furnaces using $NaCl-KCl-Na_3AlF$ flux. The materials were siliconized, sintered, and nitride-bonded SiC (Reference 23) ceramics as shown in Table 16. The exposure conditions were approximately 1100°C (2000°F) for 1318 and 2448 h with numerous thermal cycles to lower temperatures. The materials were significantly corroded during this exposure.

Projected surface recession rates ranged from about 3 mm/y for Coors' siliconized SiC and Carborundum's sintered-α to about 20 mm/y for Norton's nitride-bonded SiC. Norton's siliconized SiC had an intermediate recession rate. Corrosion mechanisms were not determined, but the authors suggest that volatilization of silicon halides occurred after the normally protective silica film was destroyed by halide species.

TABLE 16. RECESSION RATES OF SIC CERAMICS EXPOSED TO FLUE GASES IN AN ALUMINUM REMELT FURNACE AT 1100°C

Material	Designation	Manufacturer	Recession Rate[a] mm/y	Recession Rate[a] in./y
Siliconized	NC-430	Norton Co.	9.7	0.38
Siliconized	RSSC	Coors Porcelain Co.	3.3	0.13
Siliconized	Hexoloy KX-01	Carborundum Co.	--[b]	--[b]
Sintered	Hexoloy SA	Carborundum Co.	3.1-4.2	0.12-0.17
Nitride-bonded	Cryston	Norton Co.	21.4	0.84

a. Recession rate projected from 2448 h exposure except Cryston 1318 h.
b. Not measured.

These results contrast significantly with the previous results in which SiC ceramics exhibited no surface recession during exposure in an aluminum remelt furnace. This means that corrosion results are specific to particular furnaces of the same type, and that results cannot be translated from one to another.

Furthermore, the results indicate a need for characterizing flue gases and for understanding corrosion mechanisms associated with specific corrodents.

Glass Furnace Exposure--A corrosion test was conducted in a soda-lime glass furnace to compare the performance of candidate materials exposed to flue gases containing species from the glass batch.[24] Materials in the test, shown in Table 17, were silicon carbide ceramics, high purity alumina, aluminosilicates, cordierite, and silicon nitride. These materials were exposed for periods up to 116 days at temperatures of 1150 to 1550°C. The glass batch composition, shown in Table 18, contained large

TABLE 17. CERAMIC MATERIALS EXPOSED TO FLUE GASES IN A SODA-LIME FURNACE GLASS

Material	Designation	Manufacturer
Sintered SiC	Sintered-α	Carborundum Company
Recrystallized SiC	NC-400	Norton Company
Reaction sintered SiC	NC-430	Norton Company
Reaction sintered SiC	KT	Carborundum Company
Alumina	Vistal	Coors Porcelain Company
Alumina	AD-998	Coors Porcelain Company
Aluminosilicate	AD-94	Coors Porcelain Company
Aluminosilicate	AD-85	Coors Porcelain Company
Mullite	Mullite	Coors Porcelain Company
Cordierite	CD-1	Coors Porcelain Company
Silicon Nitride	RBN	AiResearch Manufacturing Company

TABLE 18. BATCH COMPOSITION IN GLASS FURNACE DURING RECUPERATOR MATERIALS TESTING

Material	Charge wt %	Equivalent type	Oxide in charge content, wt %
SiO_2	50.6	SiO_2	50.6
$CaCO_3$	14.5	CaO	8.1
Na_2CO_3	17.7	Na_2O	10.4
Dolomite	7.2	CaO	2.2
		MgO	1.6
Feldspar	8.0	Al_2O_3	1.5
$BaCO_3$	0.6	BaO	0.4
Cryolite[a]	0.8	Al_2O_3	0.4
		Na_2O	0.2
Borax	0.6	B_2O_3	0.4
		Na_2O	0.2

a. Also contains fluorine.

concentrations of SiO_2, CaO, and Na_2O (equivalent oxides). All exposed materials had a glassy surface coating containing constituents of the glass batch. In some cases, as will be described, significant corrosion was evidenced by surface recession or microstructural alteration.

o   Sintered-α SiC, 1225 to 1550°C. Significant material loss occurred. Surface recession rate was highest (6.3 μm/h) at 1225°C compared to only 1.1 μm/h at 1550°C, which indicates that corroding species were more concentrated at the lower temperatures.

o   Recrystallized SiC, 1400°C. Surface recession was observed, but not measured. An amorphous surface coating penetrated into the extensive open pores in the material.

o   Reaction sintered SiC, 1225 to 1550°C. Significant material loss occurred at all temperatures, but was highest at the lower temperatures in a manner similar to the attack on sintered-α. The attack on the free silicon phase, however, increased with increasing temperature as evidenced by disappearance of the silicon and replacement with glass or porosity.

o   Alumina (Vistal), 1150 to 1450°C. This material had the highest initial purity and density of the two aluminas tested. Some sagging occurred at the highest temperature and surface recession at the lower temperatures. Grain boundary penetration was slight compared to the other alumina and the aluminosilicates.

o   Alumina (AD-998), 1450°C. This material nominally contains 99.8% Al_2O_3. Porosity near the surface was filled with a glassy material that deposited on the surface, but no significant corrosion was observed.

o   Aluminosilicates (AD-85 and AD-94), 1150 to 1400°C. These
    materials contain 85 and 94 wt % Al_2O_3, respectively
    (equivalent oxide concentration). Sagging occurred at the higher
    temperatures, surface recession at lower temperatures. Extensive
    penetration of a surface coating into the materials resulted in
    altered microstructures.

o   Mullite, 1150 to 1450°C. This material nominally consists of
    72 wt% Al_2O_3 and 28 wt% SiO_2, equivalent oxides.
    Significant deformation occurred at higher temperatures.
    Penetration of the surface coating and microstructural alteration
    occurred at all temperatures.

o   Cordierite, 1150 to 1425°C. The ideal composition of this
    material is $2MgO \cdot 2Al_2O_3 \cdot 5SiO_2$. The melting point is
    approximately 1465°C. In this test, material exposed at 1425°C
    was not self-supporting, and was lost. Apparently 1425°C is
    above the temperature stability range. Significant penetration
    by the glassy surface layer occurred in material exposed at lower
    temperatures, resulting in extensive microstructural variations
    including new phases and porosity.

o   Silicon nitride (Si_3N_4), 1225 to 1350°C. This material was
    embrittled by exposure at 1350°C, and exhibited significant sur-
    face recession at 1225°C.

The results of this test showed that materials relying on either a
protective SiO_2 film (SiC or Si_3N_4) or an extensive silicate bonding
phase (cordierite, mullite, aluminosilicates) were severely corroded by the
glass batch species. The corrosion behavior of SiC, Si_3N_4, and oxide
ceramics containing substantial amounts of silica, therefore, appears to be
unacceptable for glass furnaces. The high-purity aluminas were the most
corrosion-resistant materials in this test. These materials might have
been operating near their upper temperature limit because some deformation
occurred. Of course, materials in this test did not operate as heat

exchanger elements and thus did not benefit from design based on mechanical property considerations or the cooling effect of air flow on one side of the flue gas/air boundary.

Residual Oil Combustion Gases Exposure--A corrosion test of candidate ceramic materials was conducted in the Ceramic Recuperator Analysis Facility (CRAF) at ORNL under controlled conditions (Reference 17). The materials, which included both SiC and oxide ceramics, were tested as tubular heat exchanger elements. Flue gases flowed over the outer surfaces while air to be heated flowed through the tubes. The outer surfaces were maintained at 1200°C (at the midpoint) during an exposure of 500 h which included six thermal cycles to lower temperatures. The fuel was No. 6 residual oil which contained Na, Ni, P, and Si as major impurities along with the usual high S content. The materials involved in this test are shown in Table 19.

Active corrosion processes were observed for the SiC ceramics. Nodules and deposits consisting of cristobalite, tridymite, and silicate glass containing Na, Ca, Al, V, Fe, and Ni formed on the upstream side of the tubes during the test exposure, probably as a result of impingement and/or condensation of fuel oil impurities. Active oxidation of silicon carbide grains and formation of gas bubbles at the SiC interface with the nodules were observed. Iron and nickel from fuel oil impurities reacted with free silicon in the siliconized SiC to produce low-melting alloys, which might degrade high-temperature mechanical properties. Large SiC grains, which might result in lower mechanical strength, were observed in exposed sintered-α SiC. The nodules or deposits on SiC ceramics were molten during the exposure.

Some oxide-based ceramic materials such as cordierite, mullite, and zirconia-mullite cracked extensively during the exposure. These oxide ceramics are considered to be of secondary interest for applications involving thermal cycling. Microcracks were also observed in exposed CVD and CVD-coated SiC. Internal cracks were observed by radiography in the exposed AD-998 alumina tube. A thin layer of iron-nickel aluminate solid formed on the outer surfaces of alumina (AD-998), probably as a result of reactions

TABLE 19. CERAMIC MATERIALS EXPOSED TO RESIDUAL OIL COMBUSTION GASES IN CRAF TEST

Material	Designation	Manufacturer
Siliconized SiC	Refel	Pure Carbon Co.
Siliconized SiC	KT	Carborundum Co.
Siliconized SiC	NC 430	Norton Co.
Sintered SiC	Sintered-α	Carborundum Co.
High-purity Al_2O_3	AD-998	Coors Porcelain Co.
Chemical vapor-deposited SiC	CVD SiC	Deposits and Composites
CVD SiC on NC 400 SiC	CVD SiC on SiC	Deposits and Composites
Oxide-bonded SiC	Carbofrax-M	Carborundum Co.
Oxide-bonded SiC	Carbofrax-A	Carborundum Co.
Si_3N_4-bonded SiC	Refrax-20	Carborundum Co.
Mullite	Mullite	Coors Porcelain Co.
Cordierite	MAS 8200	GTE Sylvania
Zirconia-mullite	Zirmul	GTE Sylvania
Cordierite	MAS 3400	GTE Sylvania

with fuel oil impurities such as iron, nickel, and aluminum. This protective aluminate layer might act as a barrier to further corrosion via reactions with flue gas impurities. Grain growth was observed in the outer surface area of the upstream side of the exposed AD-998 alumina tube.

Helium permeability of some of the materials increased as a result of exposure to residual oil combustion products. Significant increases ranging from a factor of 5 to several orders of magnitude in helium leakage rate through the post exposure tube wall were observed in siliconized SiC, sintered-α SiC, CVD SiC, CVD coated SiC, and high-purity alumina ceramics. This observation indicates the need of a critical determination of the gas leakage rate of state-of-the-art structural ceramics for use in high-temperature heat exchangers in closed-cycle energy conversion systems in which high-pressure helium or other gases may be used as working fluids.

Room-temperature postexposure C-ring fracture strengths of sintered-α SiC, CVD SiC, CVD SiC on SiC, and high-purity alumina were significantly lower than values for as-received materials. The C-ring fracture strengths at room temperature of exposed siliconized SiC and oxide- and nitride-bonded SiC increased slightly, probably because of the blunting of surface flaws and the presence of surface nodules and deposits. Strength degradation might occur at high temperatures. For example, the strength of siliconized SiC might be affected by the low-melting Fe-Ni-Si phase composed of Si originally present in the material and Fe and Ni impurities in the flue gases. High temperature strength measurements are needed to investigate this possibility. (C-ring fracture strength data is shown in Table 20.)

The coefficient of thermal expansion (CTE) of sintered-α SiC, Refel SiC, CVD SiC on SiC, Carbofrax-M, MAS 8200, and zirmul increased. On the other hand, the CTE of MAS 8400 and mullite decreased slightly and that of KT SiC, NC 430 SiC, CVD SiC, Carbofrax-A, Refrax-20, and AD-998 Al_2O_3 remained constant.

Coal Oil Mixture Combustion Exposure--A corrosion test of ceramic materials in residual oil combustion products in the CRAF was described in the previous section. Two additional corrosion tests of ceramic materials in the CRAF used coal oil mixture (COM) fuels.[25,26] The tubular materials were tested as heat exchanger elements with flue gases flowing on the outside and air on the inside. In the first test the temperature of the outer surface of the tubes was maintained at about 1240°C for 496 h using COM with acidic ash (base/acid ratio of 0.29). The second test conducted at about 1240°C for 240 h used COM with basic ash (base/acid ratio of 1.14). The materials included in these tests are shown in Table 21.

Results of Exposure to Acidic Ash--The 496-h exposure to the combustion products of the acidic-ash coal resulted in the formation of thick deposits of coal slag on all tubes. The strong interfacial bonds formed between the coal slag and both the sialon and Al_2O_3 tubes were apparently responsible for the extensive cracking observed in these two materials. This fracture problem indicates that sialon and Al_2O_3 materials are not suitable for highly fouling environments containing acidic coal ash constituents. Conversely, all the silicon carbide tubes including KT SiC, sintered-α, and CVD SiC survived the exposure without any major material degradation. From this observation and the results of a

TABLE 20. C-RING FRACTURE STRENGTHS OF CERAMIC MATERIALS EXPOSED TO RESIDUAL OIL COMBUSTION GASES

Type of Material	Designation	Manufacturer	C-Ring Fracture Strength, MPa[c]		Change %
			As-Received	Exposed	
Sintered SiC	Sintered-α	Carborundum Company	243±27	169±37	-30
Reaction sintered SiC	KT	Carborundum Company	116±17	162±12	39
Siliconized SiC	NC-430	Norton Company	115±22	172±26	50
Reaction Sintered SiC	Refel	Pure Carbon Company	121±16	174±14	43
CVD SiC	CVD SiC	Deposits and Composites	217±74	229±48	6
CVD SiC on NC-400 SiC	CVD SiC on SiC	Deposits and Composites	150±20	99±41	-34
Nitride-bonded SiC	Refrax 20	Carborundum Company	19±1	42±10	160
Oxide-bonded SiC	Carbofrax-A	Carborundum Company	12±5	25±2	108
Oxide-bonded SiC	Carbofrax-M	Carborundum Company	14±3	16±3	12
Alumina	AD-998	Coors Porcelain Co.	194±19	126±10	-35

a. Source: Reference 17.

b. 500 h at 1200°C with six thermal cycles; major oil impurities (ppm) were: Na(76), Ni(10), Si(47), V(83), P(46), and 2.7% S.

c. Average strength at ~25°C of three or more specimens.

TABLE 21. CERAMIC MATERIALS EXPOSED TO COMBUSTION PRODUCTS OF COAL OIL MIXTURES

Material	Designation	Manufacturer	Test[a]
Reaction sintered SiC	KT	The Carborundum Co.	A, B
Sintered SiC	Sintered-α	The Carborundum Co.	A, B
Chemical vapor deposited SiC	CVD	Deposits and Composites	A, B
Alumina (Al_2O_3)	AD-998	Coors Porcelain Co.	A, B
Sialon	GE-128 Ge-129, -130	General Electric Co.	A, B B
Reaction sintered SiC	SC-2	Coors Porcelain Co.	B
Siliconized SiC	NC-430	Norton Co.	B
Chemical vapor-reacted SiC	CVR	Syntax Corporation	B

a. A - acidic coal ash; B - basic coal ash.

previous CRAF test using No. 6 oil, silicon carbide ceramics appear to be good candidates when either No. 6 oil or acidic coal ash fuel impurities are present in the environment.

With the exception of the KT SiC, ceramographic and SEM analyses revealed no significant changes in the microstructures of the silicon carbide materials following the 496-h exposure. A few isolated corrosion pits formed along the upstream surface of the KT SiC; however, the total amount of surface corrosion was still quite limited. The microstructural analyses of the sialon and Al_2O_3 tubes revealed a 150-μm-thick oxida- tion layer at the sialon-slag interface and a 6- to 12-μm-thick aluminosilicate layer at the alumina-slag interface. The strong interfacial bonding coupled with the thermal expansion mismatch between the slag and both ceramic materials might have caused large tensile stresses in the respective tubular elements upon cooldown from the 1240°C test temperature. These stresses would have been sufficient to cause fracture in the alumina tubes.

The helium permeabilities of the surviving exposed silicon carbide tubes were not significantly different from the respective values measured for the as-received materials at each pressure differential.

A fairly large increase in the room-temperature C-ring strength of the KT SiC occurred during the test exposure. This increase was apparently related to high-temperature flaw blunting processes, which decreased the severity of the surface flaws. These processes were probably facilitated by deformation of the silicon phase, which is consistent with the absence of significant strength increases for either CVD or sintered-α SiC. Both these materials contain little or no free silicon. The strength changes in Al_2O_3 or sialon could not be ascertained because of the extensive cracking of the exposed materials. (C-ring fracture strength data are shown in Table 22.)

<u>Results of Exposure to Basic Ash</u>--The 240-h exposure of the ceramic tubes to the combustion products of the basic-ash coal resulted in the formation of relatively thin deposits of coal slag on all the tubes. The basic ash caused extensive and nonuniform wall thinning of materials especially on the upstream side of the tubes. Siliconized SiC tubes had upstream wall losses of approximately 20 to 33%, while sintered-α SiC had a value of about 16%. The CVD SiC tube had the lowest value of 13%. The maximum Al_2O_3 tube wall loss was moderate, being around 15%, while the three types of sialon ceramic tubes had maximum wall thickness losses of about 20% to 78%. The observed material losses from the upstream side of the tubes of this wide variety of ceramic materials demonstrate that under the high-temperature fluid basic coal-slag conditions employed in this exposure, none of these materials have sufficient corrosion resistance to serve for long time periods as recuperator elements unless very thick sections are employed.

TABLE 22. C-RING FRACTURE STRENGTHS OF CERAMIC MATERIALS EXPOSED TO COAL OIL-MIXTURE COMBUSTION PRODUCTS CONTAINING EITHER ACIDIC OR BASIC COAL ASH

Material	Designation	Manufacturer	C-Ring Fracture Strength, MPa [b]					
			Acidic Coal Ash [c,d]			Basic Coal Ash [e]		
			As-Received	Exposed	Change, %	As-Received	Exposed	Change, %
Reaction sintered SiC	KT	Carborundum Company	179±8	199±15	11	203±13	180±15 [f]	-47
Reaction sintered SiC	SC-2	Coors Porcelain Co.	[h]			179±8	72±229	-60
Siliconized SiC	NC-430	Norton Company	[h]			239±23	93±52	-61
Sintered SiC	Sintered-α	Carborundum Company	305±52	292±24	-4	174±18	109±23	-38
Sialon	GE-128	General Electric Co.	93±10	117±10	26	264±30	255±47	-3
Alumina	AD-998	Coors Porcelain Company	147±19			[i]		
Chemical vapor deposited SiC	CVD	Deposits and Composites	347±54	335±41	-3	241±22	174±45	-28
Chemical vapor reacted SiC	CVR	Syntax Corporation	[h]			85±3	71±20	-16

a. Source: References 25 and 26.

b. At room temperature.

c. 500 h at 1240°C, six thermal cycles; major elements in flue gases were Na, Mg, Al, Si, S, K, Ca, Ti, V, Cr, Mn, Fe, Ni, Zn.

d. 11-14 specimens for each material except only 5 Sialon specimens.

e. 240 h at 1240°C, two thermal cycles; major elements in flue gases were Mg, Al, Si, S, Ca, V, Fe.

f. Wall thickness 12.7 mm.

g. Wall thickness 6.4 mm.

h. Not included in this test.

i. Not tested because of irregular wall thickness.

j. Extensively cracked.

The SiC ceramics corroded via formation of phases at the ceramic-slag interface which were tentatively identified as iron and nickel silicides, with iron and nickel coming from the slag and the silicon presumably coming from the SiC. These intermetallic type phases then become detached from the SiC surface and are transported into the fluid slag, whereupon the smaller particles react with the slag and disappear. Development of corrosion pits and channels in the outer surface of the tubes resulted in substantial reductions in fracture strengths at room temperature. Sintered-α SiC exhibited the lowest reduction (-3%) in average fracture strength of all the SiC ceramics studied. Siliconized SiC specimens showed strength reductions from approximately 28 to 60%. The highly heterogeneous CVR SiC exhibited a modest strength reduction of 16%; however, the initial strength of this ceramic was the lowest of all the materials included in this experiment. (C-ring fracture strength data are given in Table 22.)

The AD 998 Al_2O_3 ceramic experienced wall thinning by a corrosion process in the coal slag via a mechanism quite different from that observed for the SiC ceramics. The Al_2O_3 ceramic-slag interface became uniformly coated with an Fe-Ni aluminate. This primary reaction product phase then reacted with the fluid slag, resulting in continuous transport of aluminum from the Al_2O_3 into the slag. This process produced a microscopically rough surface on the Al_2O_3 which contributed to a fracture strength reduction of about 28% due to slag exposure.

Exposure to the hot coal slag increased the room temperature helium permeability of all the surviving tubes by about one order of magnitude. The thermal expansions of the SiC and Al_2O_3 based ceramics were essentially unchanged by long-term exposure to the fluid coal slag.

<u>Summary</u>--The corrosion results presented indicate that the effects of flue gas contaminants on mechanical properties depends upon the type of contaminants. The most severe strength degradation (measured at room temperature) was caused by basic coal ash, which contains relatively large concentrations of Ca, Fe, K, Mg, and Na. Acidic ash, which contains relatively large concentrations of Al, Si, and Ti, caused very little

strength reduction. Additional, more detailed information is needed on
flue gas composition and the effects of flue gas contaminants on high temperature mechanical properties before the behavior of ceramic recuperator
materials can be adequately predicted.

Conclusions

This assessment reveals that the data base for physical and mechanical
properties of structural ceramics is quite incomplete, especially at temperatures of interest for recuperators. Although a large data base
typically exists for metallic alloys at elevated temperatures, lack of
information hampers the selection of structural ceramics. A more inclusive
data base will probably become available as particular structural ceramics
become mature, that is, optimized with respect to constituent materials and
fabrication methods. Meanwhile, the user must develop the parts of the
data base that are critical to specific applications. The minimum requirement for materials selection and recuperator design is the high temperature
tensile or flexural strength with Weibull modulus. This information is
needed to avoid excessively thick load-bearing sections in tubes or other
components. Static fatigue data at high temperatures are needed for conservative design and for lifetime prediction. Recuperator design would be
greatly facilitated by the availability of other high temperature properties including thermal conductivity, thermal shock resistance, Young's modulus, fracture toughness, Poisson's ratio, and creep strength. In addition, the effects of corrosion on high temperature mechanical properties
are essentially unknown, and are needed for conservative design. Obtaining
this information, however, is complicated by the possibly different corrosion potentials of various industrial flue gases.

Improved Materials

Improved Processes

The wide distribution in material properties, especially as it affects the expected durability, may be attributed to the brittle behavior of ceramics and to the impurities in the starting powder. In light of this, techniques for improving currently available ceramics through increased toughening and improved powder processing must be addressed. Work is, in fact, under way to improve materials and processes including fiber or particle toughened ceramics and synthesis of raw materials. Some of this work is described in the following paragraphs.

Synthesis of SiC Powder. Silicon carbide ceramics are currently fabricated from powder that is synthesized by several methods. Several major manufacturers use SiC powder synthesized by the Acheson process.[27] This process, which has been in use for about 100 years, provides SiC for both abrasive and structural ceramic applications. A flow sheet for the process is shown in Figure 11. A mixture of silica, carbon, sawdust, and salt are heated by an electric arc initially to about 2700°C to initiate carbothermic reduction and subsequently to 2000°C or higher until the reaction is completed. The product is primarily polycrystalline α-SiC. Extensive crushing, grinding, sizing, leaching, and washing is required to prepare powder for fabrication of structural ceramics. These additional processing steps, which are necessary to prepare suitable powder from the Acheson product, results in a significant increase in cost while generating powder of varying batch to batch properties.

Other commercial processes for obtaining SiC powder involve synthesis by vapor-phase techniques. Vapor phase techniques lead to very fine unaggregated powders, usually of spherical geometry. Particle size can be controlled by manipulating concentrations and temperatures with typical sizes ranging between 2 and 200 nm. The major disadvantages result from difficulties in collecting the final product powder and limited production

Ceramic Materials for High Temperature Application   287

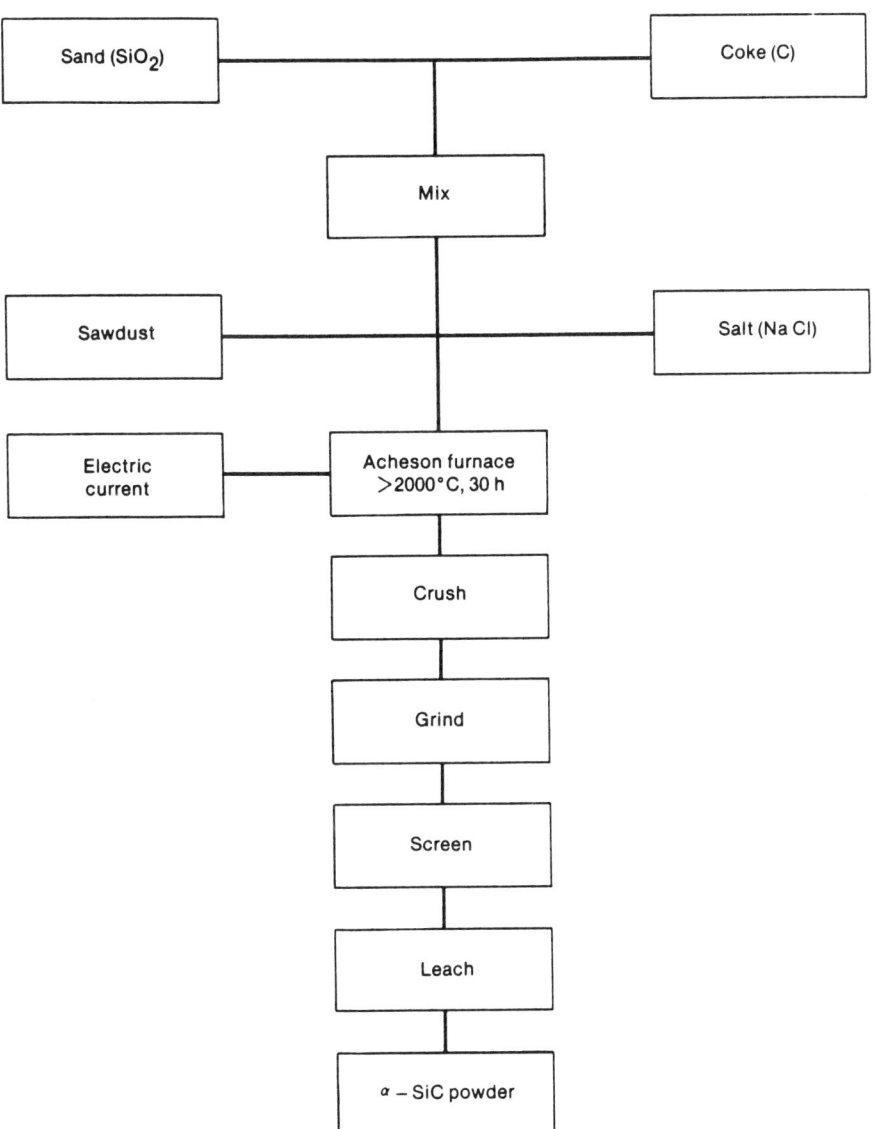

Figure 11. Flow sheet for the Acheson synthesis process.

facilities. Vapor phase techniques for producing carbides include vapor condensation methods, vapor decomposition and vapor-vapor reactions.

An alternative is needed to the Acheson and vapor reaction processes for producing relatively inexpensive SiC powder. The powder produced by the alternative process must have all of the physical and chemical attributes required to fabricate high quality SiC ceramics. Several powder synthesis processes presently being developed are described below.

The high-temperature rotary furnace process, which is a refinement of the Acheson process, is being investigated at ORNL.[27,28] In this process the reaction of silica and carbon to form ultrafine SiC particles is controlled under better heat transfer conditions than in the Acheson process. The process shown by the flow sheet in Figure 12 is continuous. The reactants, silica and carbon, are loaded into one end of the slightly tilted rotary kiln, reacted in the hot zone, and discharged at the exit end of the kiln. The constant movement of the reactants and their direct contact with the tube wall in the rotary furnace provide convective and conductive heat transfer to the reaction interface between silica and carbon and permit SiC to form at a maximum tube wall temperature much less than the 2000°C required in the Acheson process. Because of the improved heat transfer and lower synthesis temperature, less energy is required for synthesis in the rotary furnace. Using proper feed materials, this process has the potential advantage of continuous synthesis of high-purity sinterable SiC powder at a low cost.

Early results demonstrated the feasibility of synthesizing SiC powder by this process; however, much more work is required to optimize the product in terms of process variables. The first powder prepared in the rotary furnace had a low impurity content, a submicron particle size, and was fabricable by hot pressing to 96% of theoretical density. Development of this process is continuing.

The rotary furnace process was first investigated by ORNL and Eagle Picher Company, Energy Products Division, Oak Ridge, Tennessee. Process feasibility was established. Subsequently, Advanced Refractory

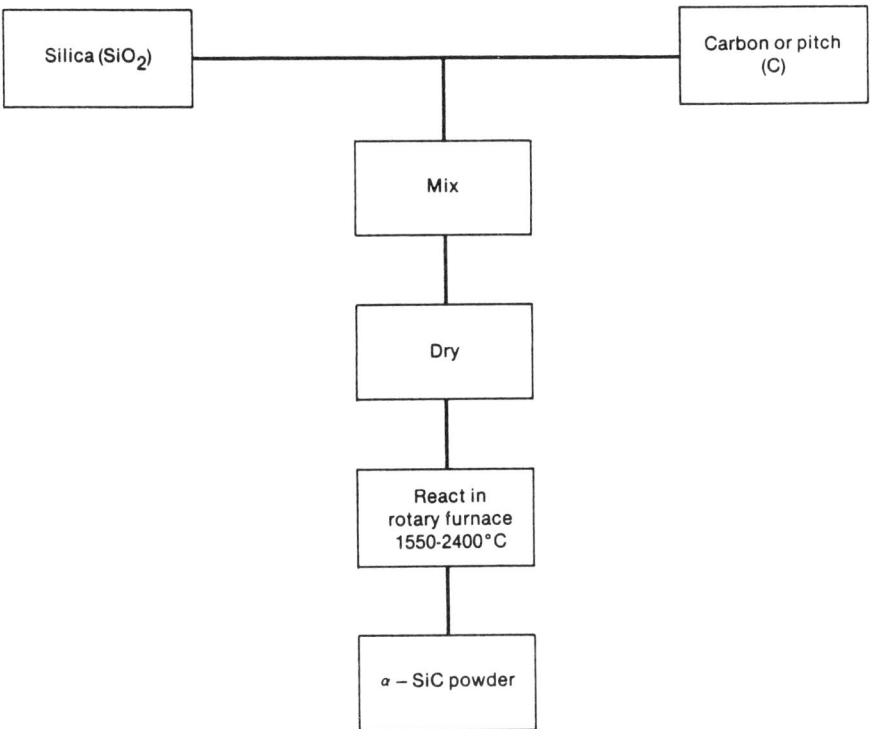

Figure 12. Flow sheet for rotary furnace SiC powder synthesis process.

Technologies, Inc. of Buffalo, New York, acquired the Eagle Picher Company facilities, and is using the process to manufacture grinding abrasives. Advanced Refractories Technologies, Inc. estimates that 50% less energy is required to make SiC in the rotary furnace compared to the Acheson process.

Gel Process. This process was initially described by Prochzaka.[29] The concept is being further developed at ORNL in parallel with the rotary furnace process described above and the polymeric precursor process described below. The flow sheet for the gel process is shown in Figure 13. Synthesis initially involves the formation of a gel of high-purity silica doped with carbonaceous species (resin, sugar, etc.) followed by dehydration and pyrolysis of the gel to form a finely divided mixture of silica and carbon. Continued heating results in reaction of the silica and carbon to form β-SiC above 1550°C. The advantage of gel processing is intimate mixing of the reactants, including dopants, on an atomic scale. Two types of gel process (Figure 13) are being investigated. These gel processes involve a mixture of colloidal silica, petroleum pitch, and toluene in Example 1 or a polymerization of alkoxysilane and resin solutions in Example 2. The gel mixtures are dried, pyrolyzed, and reacted in a stationary or rotary furnace to produce sinterable SiC powders.

The fabricability of initial powder prepared by the gel process has been determined. Powder has been sintered to 95% and hot pressed to 99% of theoretical density. These early results are very encouraging. Both synthesis and fabrication studies are in progress.

Polymeric Precursor Process. This process is also being investigated at ORNL. In recent years thermal setting polycarbosilanes, which can be pyrolyzed to SiC, have been synthesized. Two examples of the process, based on work by Yajima et al.[30] and Schilling et al.[31] are illustrated in Figure 14. The process, which can be used to prepare both SiC fibers and powders, offers potential for lower temperature processing than other synthesis methods.

Ceramic Materials for High Temperature Application   291

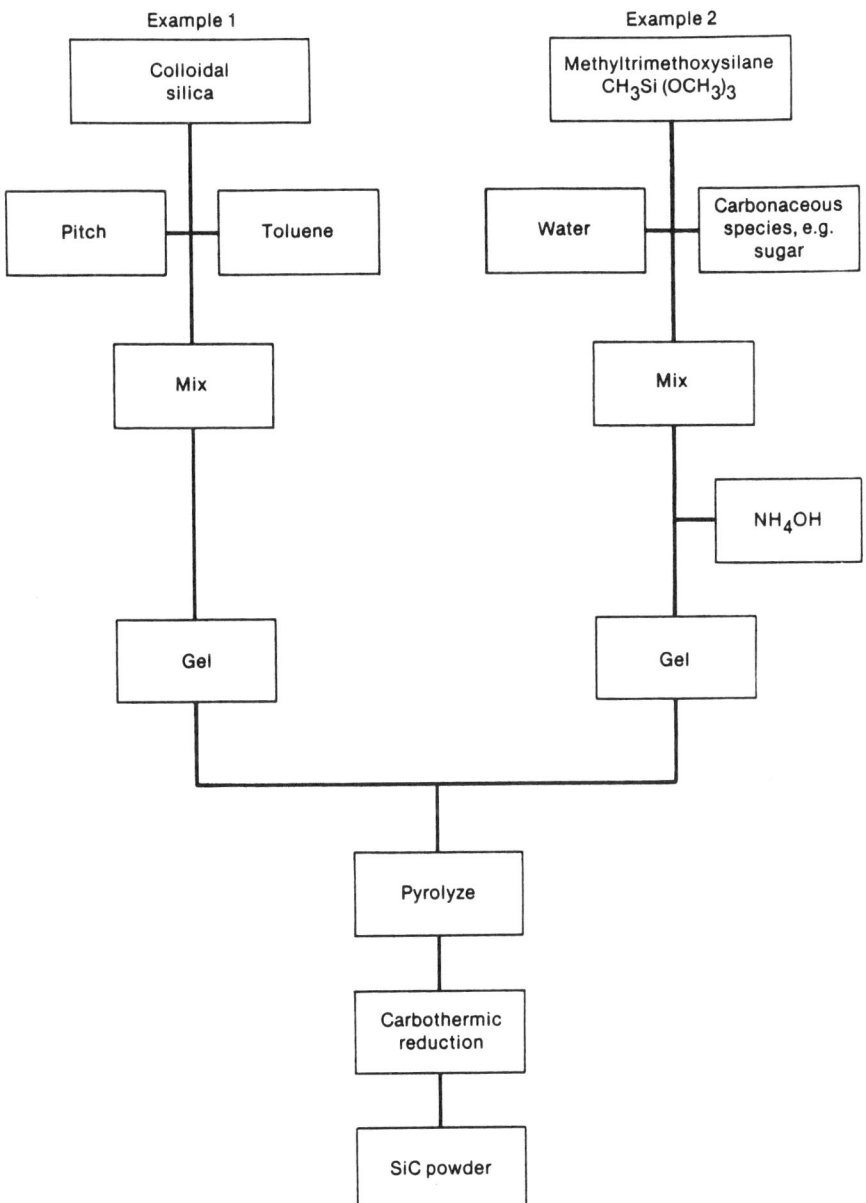

Figure 13. Flow sheet for Gel SiC powder synthesis process.

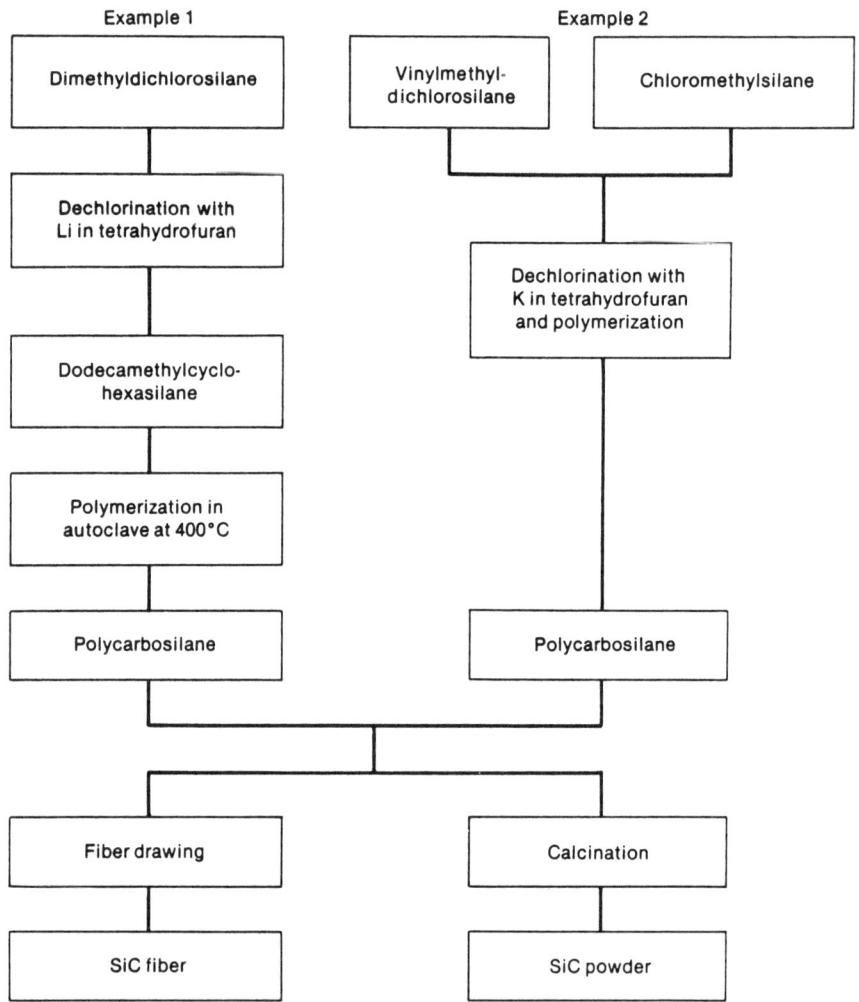

Figure 14. Flow sheet for polymeric precursor SiC powder synthesis process.

Laser-Heated Gas-Phase Reaction Process. Another SiC synthesis process being investigated at ORNL involves vapor phase reactions of SiH_4-CH_4 and SiH_4-C_2H_4 mixtures[32,33] heated by a laser. Potential advantages of the process are very small particle size, narrow particle size range, and high purity. Early results revealed no SiC formation with SiH_4-CH_4 mixtures. Powder produced with SiH_4-CH_4 mixtures were substoichiometric with respect to C, but strong Si-C bonds indicative of SiC formation were detected. This process is still in a very early stage of development.

Toughened Ceramics

Aluminum Oxide. Aluminum oxide (alumina) is expected to have good corrosion resistance in oxidizing combustion atmospheres. The apparent resistance of alumina to corrosion by a glass furnace atmosphere was discussed earlier. The fracture behavior of alumina, as in the case of other ceramic materials, is strongly influenced by the types and distribution of inherent flaws. The tensile and flexural strengths of alumina presented earlier in this section are characteristic values obtained on specially prepared specimens, e.g., by fine grinding of tensile surfaces to minimize the effect of flaws. Recuperator tubes, however, might fail at stresses substantially lower than the maximum values by propagation of flaw-induced cracks. Aluminum oxide, a strong and corrosion-resistant material, lacks toughness, which is the ability to resist premature failure by crack growth. The stresses resulting from thermal gradients during thermal cycling frequently result in catastrophic mechanical failure. Methods are needed to increase the toughness of alumina and other ceramic materials so that their fracture behavior will be more predictable.

Methods of toughening alumina are being investigated. Following the example of Prewo and Brennan,[34] who toughened glass-ceramics with SiC fibers, Becher et al.[35] are toughening alumina with SiC whiskers. Fully dense alumina containing 20 vol% SiC whiskers have been prepared by hot pressing. The stress intensity required for crack propagation in this material was approximately two times the value required for fine grained alumina showing that significant toughening was achieved. Development of this process is continuing.

Alumina has also been toughened by incorporating small amounts of finely divided metal particles.[36,37] Alumina-platinum cermets were prepared from a slurry of alumina powder and a solution of chloroplatinic acid. The slurry was evaporated to dryness, then heated in H_2 to 850°C to reduce the platinum compound to metal. The resulting material, consisting of alumina particles coated with Pt, was hot pressed at 1600°C to produce a dense body (97-99% of theoretical density). After hot pressing, the Pt was uniformly distributed throughout the alumina as fine particles. Alumina-chromium cermets were similarly prepared using $Cr(NO_3)_3 \cdot 9H_2O$ as the metal precursor.

A qualitative determination of toughening was made by quenching the cermets from 520°C into water at 80°C, which resulted in a cooling rate of ~600°C/s. Whereas alumina processed in the same manner as the cermets contained visible cracks after one or two quenches, the better cermets did not exhibit cracks after 50 quenches (Reference 36). A direct measure of fracture toughness was made using applied moment double cantilever beam specimens. Some cermets exhibited fracture toughness values significantly better than that of alumina, and comparable to the toughness of some Si_3N_4 and SiC ceramics (Reference 37). Although these results are somewhat preliminary, the feasibility of toughening alumina with SiC whiskers and metal particles has been established. Further improvements in toughening and application of the methods to other systems is expected as the investigations continue.

SiC and Si_3N_4. Recently, work was started on chemical vapor deposition of SiC and Si_3N_4 with other phases incorporated for improved strength, toughness, and wear resistance.[38] The systems being investigated include composites of SiC with either Cr, Ni, or Si_3N_4 particles and Si_3N_4 with TiN particles or fibers. If successful, this work might be extended to include oxide ceramics such as alumina, mullite, andzirconia. The feasibility of preparing some of the above composites has been established; however, properties of the materials are not yet available.

5. Design-Related Material Technologies

The ability of material property tests to obtain the necessary design information and evaluate it (including lifetime prediction) is assessed in this section. The capability of current technology to assure a given durability is also assessed.

Test methods to obtain the major mechanical property data are discussed in this subsection. The short term tensile tests, bend tests, radial compression tests, and "C" or split ring tests are discussed first. The long term time-dependent creep tests, slow crack growth tests, and corrosion tests are then discussed.

Mechanical Strength Test Methods

Tensile Test Methods

The greatest emphasis is placed upon tensile data owing to the generally accepted premise that failure in nearly all ceramic devices occurs by tensile stresses. The basic tensile test, often used for metals, is the simplest test to analyze. A summary of tensile test methods is shown in Figure 15. Several testing configurations presented by Kingery[39] and Bollard and Taggart[40] are given in Figures 16 and 17. A meaningful tensile test requires that all non-aligned (non-σ_{xx}) forces be zero and that $\sigma_{tensile}$ = Load/Area. Poisson's ratio, μ, simply becomes E_{yy}/E_{xx}; and Young's modulus, E, $\sigma_{xx}/\varepsilon_{xx}$. The elastic properties can easily be obtained via various strain gage techniques up to 1000°C. However, above 1000°C, different methods are required (References 39 and 40) such as dynamic mechanical resonance or ultrasonic techniques.

There are several drawbacks to tensile tests. Great effort must be taken to ensure proper alignment in the test apparatus. The sources of error are discussed in the literature (References 39 and 40). Design improvements, such as the "gas-bearing" device (Reference 40) minimize potential sources of measurement error. Another major drawback to this

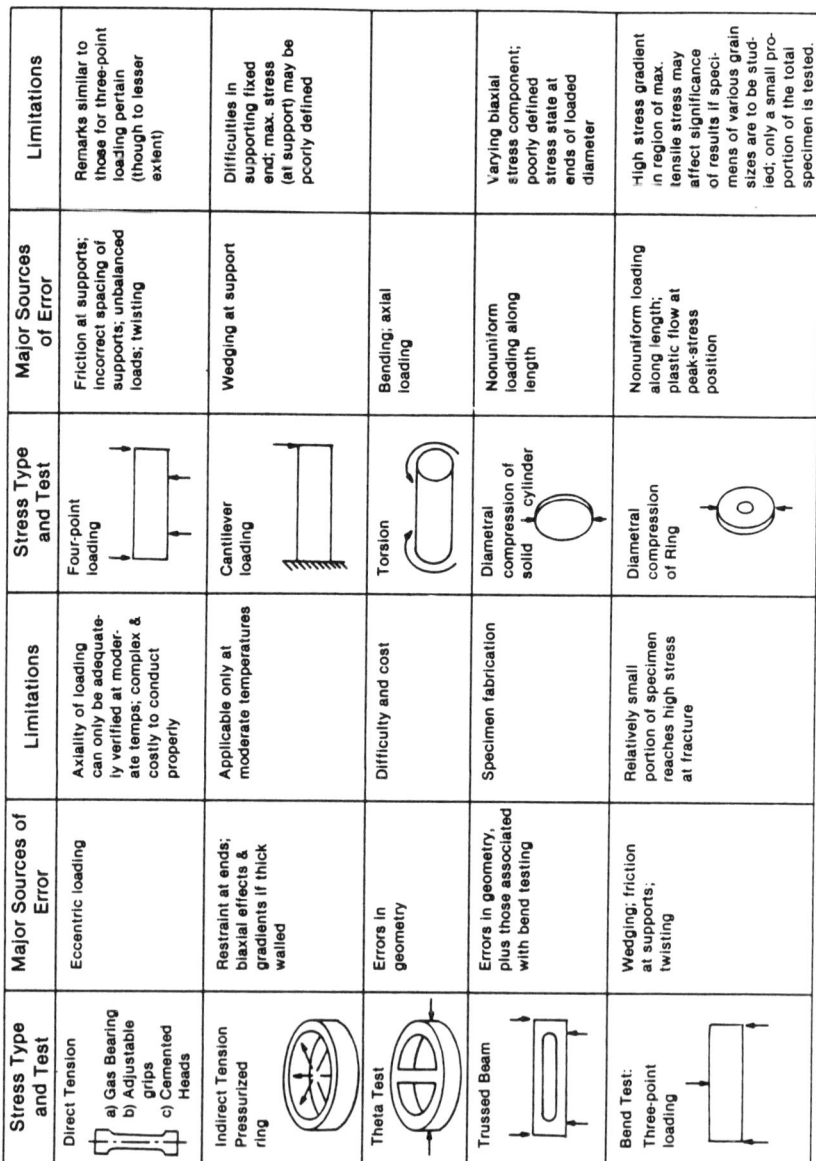

Figure 15. Summary of tensile test methods (taken from material presented at the AGARD Specialist Meeting on Brittle Materials, London, September 4, 1967. Taken from Reference 40.

Design-Related Material Technologies 297

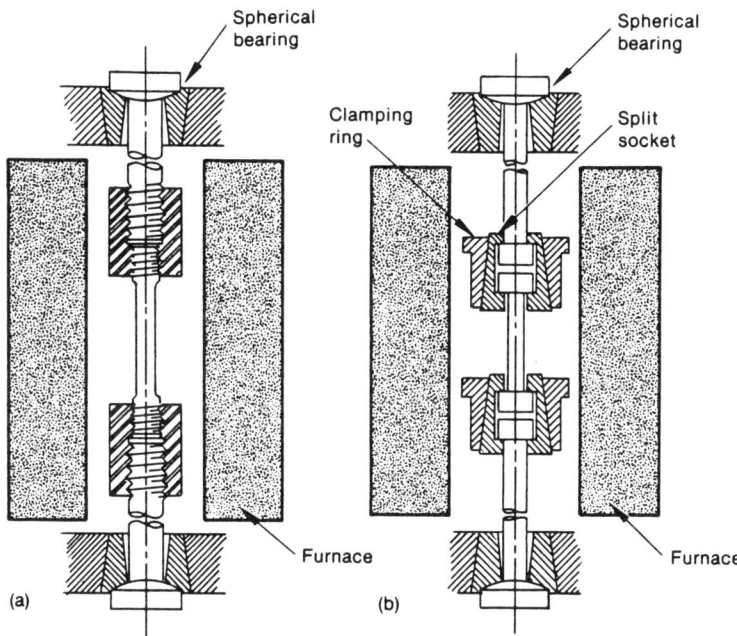

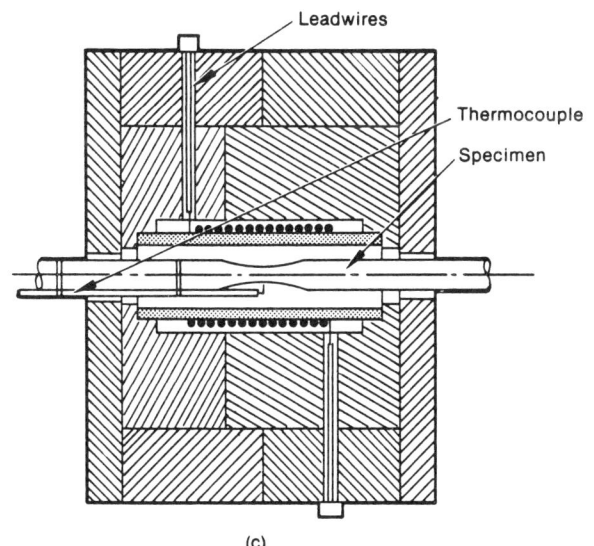

Figure 16. Various types of tensile test specimens.

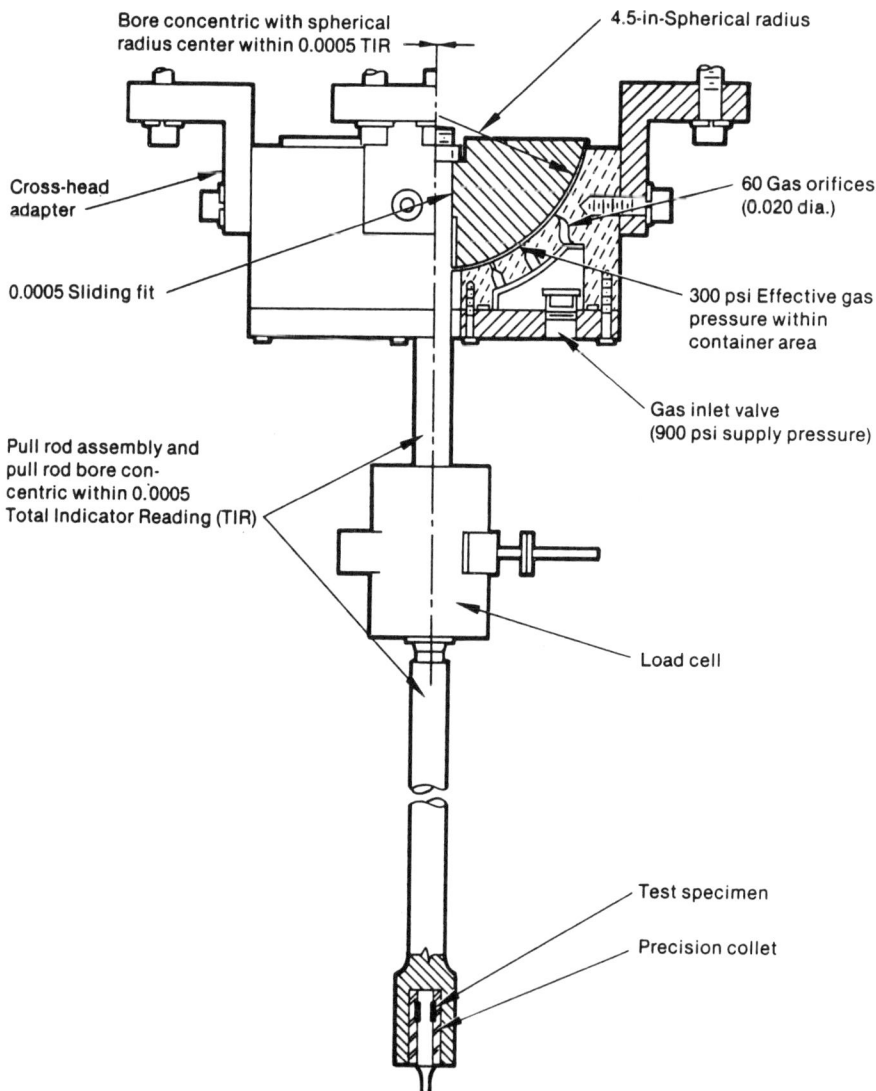

Figure 17. General layout of spherical gas bearing test rig. (Dimensions in inches.) Taken from Reference 40.

test is the difficulty and cost in machining circular test specimens to the required tolerances. This is even more of a problem with the very hard covalently bonded ceramics, such as SiC and Si_3N_4.

Bend Test Methods

The majority of strength data for various ceramic heat exchanger, adiabatic engine, and ceramic turbine programs have been obtained by modulus of rupture (or flexure) tests in which simple shapes are subjected to threeor four-point bending until fracture occurs.[41] While the solutions for fiber stresses and elastic behavior are more complicated than the simple tensile method, the relative ease in producing specimens and conducting the tests have made the bend tests extremely popular with both material researchers and product developers. The moment diagrams for the idealized loadings are given in Figure 18. Factors such as load mislocation, beam twisting, friction, local stresses, contact point tangency shift, and specimen surface conditions can introduce errors, for which there are correction factors (Reference 40).

The ability to test relatively small specimens facilitates flexural testing at high temperatures. A typical high temperature apparatus is shown in Figure 19.[42] A slightly modified version of the flexure test has been used to test tube shapes (a primary heat exchanger configuration). The fixture modification is shown in Figure 20 (Reference 42). The solutions for calculating fiber stresses and elastic properties are more complicated for this geometry, which limits its use as a materials research tool.

Radial Compression Test Methods

While it is not the intent of this section to discuss only heat exchanger components, two tests are worth mentioning that are often applied to heat exchanger tubes. The first is a radial compression test of tubular sections[43] shown schematically in Figure 21. The tube is loaded, as shown, in simple compression until failure is observed.

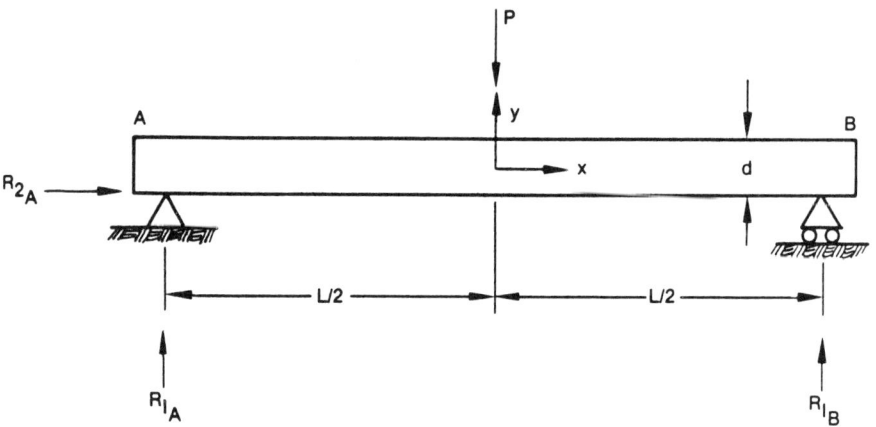

Figure 18a. Simple beam subjected to three-point loading.

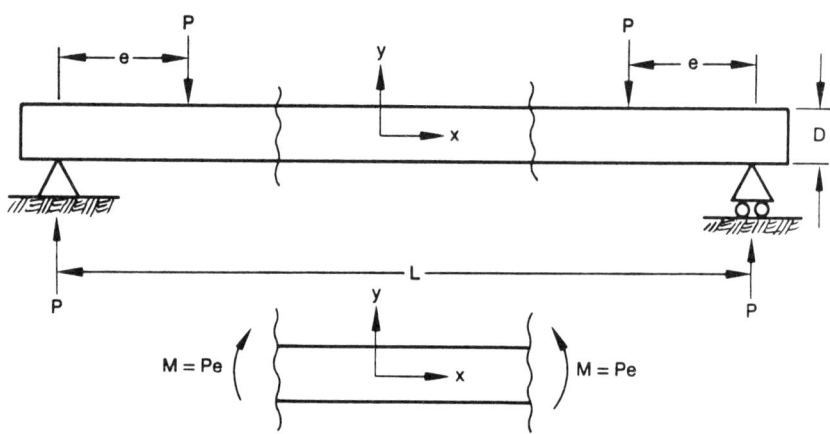

Figure 18b. Simple beam subjected to four-point loading.

Design-Related Material Technologies 301

Figure 19. Four-point bar flexure strength apparatus.

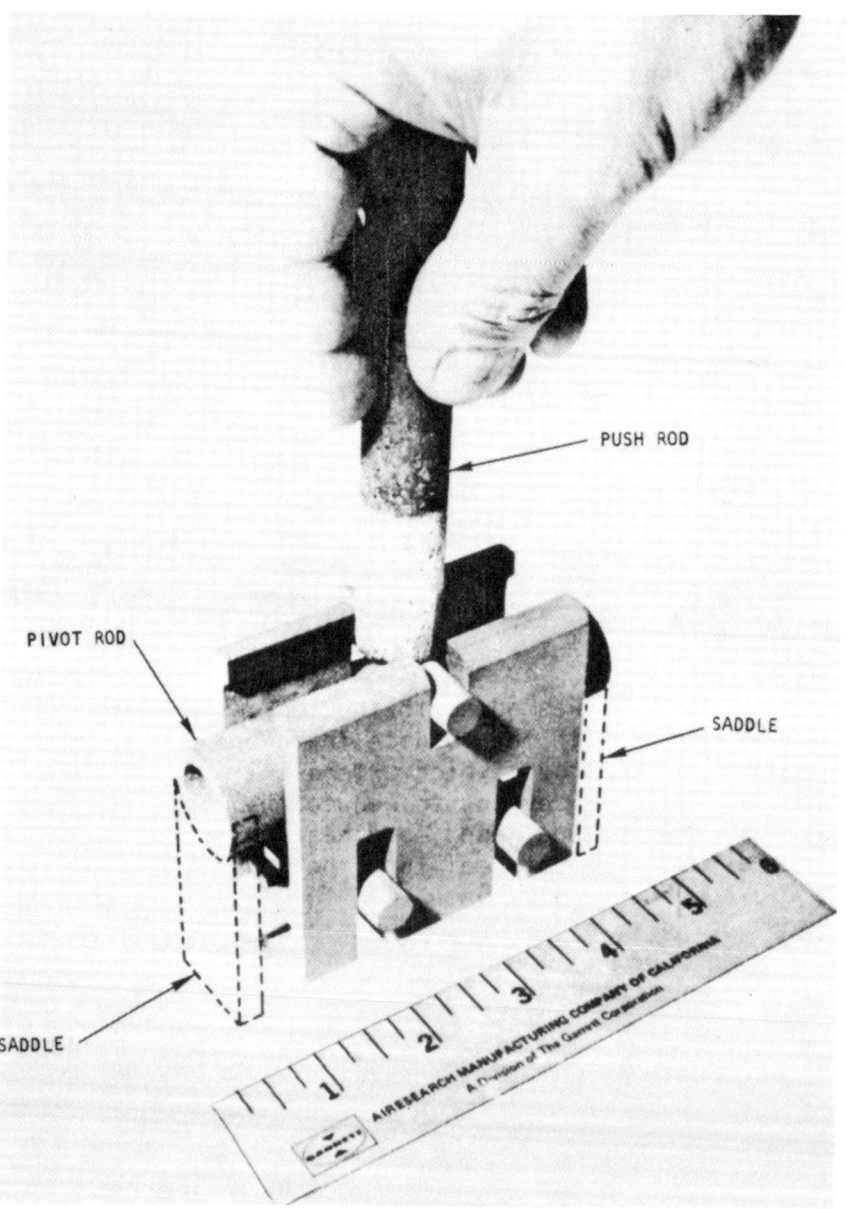

Figure 20. Fixture modifications.

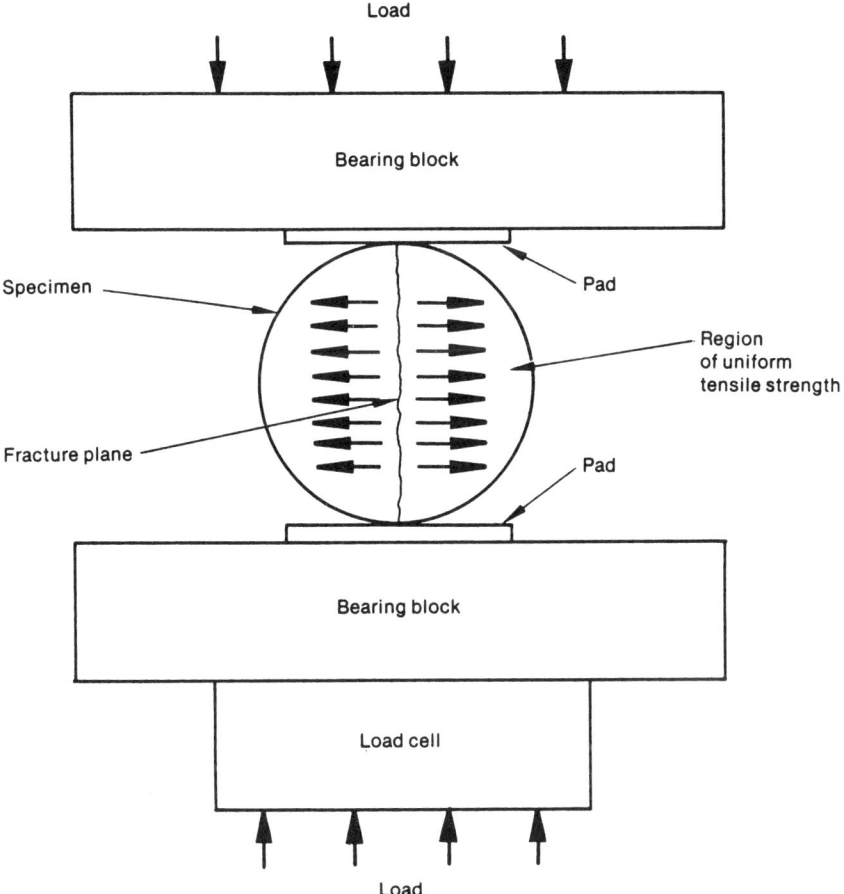

Figure 21. Radial compression test specimen and apparatus.

The main disadvantage of this test is the difficulty in machining samples. Results must be obtained, for all practical purposes, on as received materials. This may also lead to problems in applying a load uniformly to the specimen.

The second method used for solid cylindrical specimens is the splitting tensile or diametral compression method.[44,45] A schematic diagram of the specimen orientation and apparatus is given in Figure 22. The hoop strength, or tensile strength, σ, is calculated from the measured load at failure, P, by the equation

$\sigma = 2P/Dt$,

D = specimen diameter
P = load
t = specimen thickness.

A critical factor in this test is that the fracture should occur as shown in Figure 22 (vertically from pad to pad). The use of rubber or balsa wood pads help prevent localized contact fractures which would render the test invalid.

"C" or Split Ring Test Methods

A test which utilizes a "near net" heat exchanger shape is the "C" or split ring test used by Ferber and Tennery[46] to test the effects of waste stream exposure on tubular ceramic specimens. The specimen is readily obtained by sectioning tubes into rings and then removing a cross-sectional piece to produce a slit. The basic specimen geometry is illustrated in Figure 22. The fracture stress, s, is calculated from the expression:

$$s = \frac{P}{(r_o - r_i)w} \left[1 + \frac{(r_o + r_i)}{\varepsilon r_o} \frac{R - r_o}{r - R} \right]$$

where

P = the load at failure

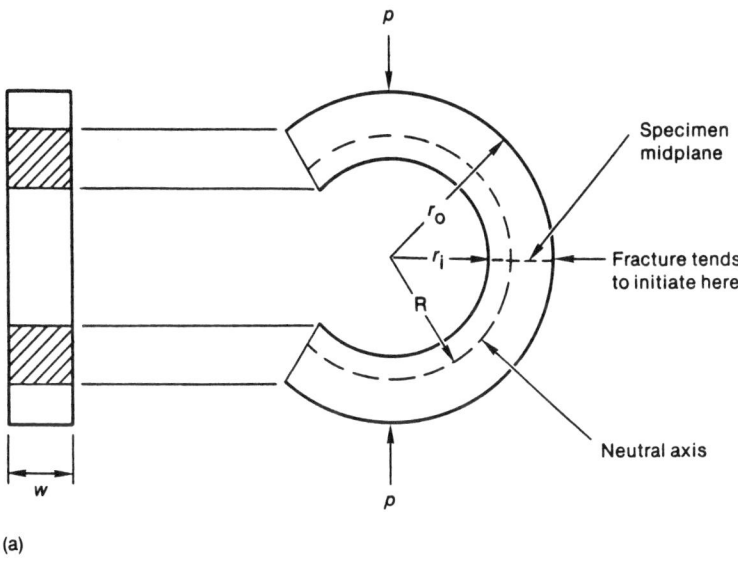

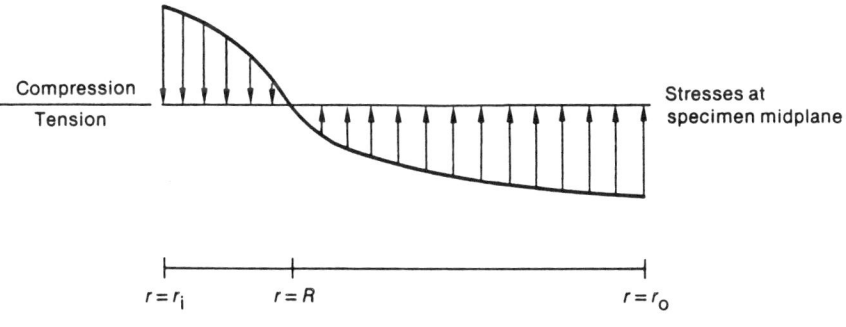

Figure 22. C-ring specimen geometry, showing the stress distribution resulting from the application of a compressive load.

$$r = \frac{r_o + r_i}{2}$$

$$R = (r_o - r_i) / (r_o/r_i)$$

A major advantage of the split-ring test is that it can be conducted on materials in their as-fabricated conditions; therefore, the microstructure is representative of actual tube material. This removes some of the uncertainty when tests are conducted on small bars etc., which may not have been fabricated in a procedure similar to tubular components. Caution, however, must be used during testing to insure that extraneous stresses are not generated that would affect the results. Adaptation of this test to the waste heat streams of interest would need to be done.

Time-Dependent Testing

Creep Test Methods

The determination of creep in any stress state requires the measurement of a material's deflection at a constant stress and temperature. The test apparatus is virtually the same as that used for strength measurements, with the exception that greater instrumentation is brought to bear in the measurement of deflection. Creep measurements can be taken in three-point or four-point bend configurations, as well as in the basic tensile mode.[47]

Creep mechanisms can be very complex depending on temperature, stress, grain size and porosity, impurities and environment. It is therefore necessary to conduct extensive creep testing in conjunction with diffusional studies, scanning electron microscopy, and transmission electron microscopy to interpret the mechanism of creep. Possible mechanisms include atomic diffusion, cavitation, and softening of a lower melting glassy phase or impurity phase.

Slow Crack Growth Test Methods

Slow crack growth can occur during constant stress, constant stressing rate, and during cyclic loading which can be considered a combination of static and dynamic conditions. Under these loading situations, a flaw or flaws can extend subcritically by one of several possible mechanisms until its size becomes critical and catastrophic failure occurs. Mechanisms of slow crack growth include stress-enhanced chemical reactions at the crack tip involving corrosive species present in the surrounding environment, internal oxidation of existing pores, and viscous flow/cavitation mechanisms in which cavity nucleation and coalescence eventually links with the main flaw and resulting in material failure.

A variety of methods have been developed to obtain slow crack growth data during static or dynamic fatigue. Many of these consist of testing fracture mechanic type specimens (see Figure 23 for examples) containing large (macroscopic) flaws introduced mechanically into the sample. Evans[48] has extensively reviewed these specimens. The applicability of each of the techniques is given in Table 23. The specimen geometries and crack tip stress intensity solutions are listed in Figure 23. As indicated by Evans (Reference 48) and Wiederhorn,[49] measurements are usually taken during constant load or constant displacement rate conditions in these large fracture mechanic type specimens. The results can be illustrated in crack velocity versus stress intensity diagrams which can be used in lifetime prediction models as will be discussed later.

A technique which has also been used successfully to monitor slow crack growth in materials during dynamic fatigue is constant stressing rate testing. A sample similar to that used in bend strength measurements is loaded in three- or four-point bend at a constant stressing rate until failure occurs. By varying the stressing rate used from test to test, a change in fracture stress with stressing rate will be observed in those materials exhibiting slow crack growth from inherent flaws. Plotting the fracture stress versus the stressing rate yields straight lines with slopes equal to 1/1+N) where N is the measure of the material's resistance to slow crack growth.

(a) Single-edge notched beam (SENB) specimen

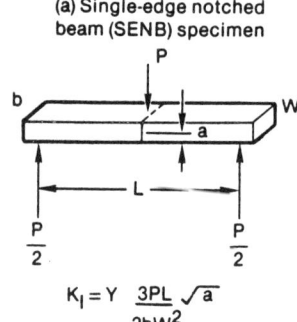

$$K_I = Y \frac{3PL}{2bW^2} \sqrt{a}$$

$$Y = 1.96 - 2.75\left(\frac{a}{W}\right) + 13.66 \left(\frac{a}{W}\right)^2 - 23.98 \left(\frac{a}{W}\right)^3 + 25.22 \left(\frac{a}{W}\right)^4$$

(b) Double torsion (DT) specimen

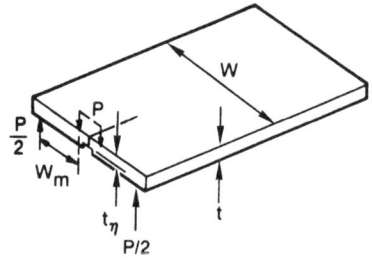

$$K_I = PW_m \left[\frac{3(1+\nu)}{Wt^3 t_\eta}\right]^{1/2}$$

(c) Constant-moment specimen

$$K_I = \frac{\text{Moment}}{\sqrt{It}}$$

I is inertia of one arm.

(D) Tapered double cantilever beam (TCB) specimen

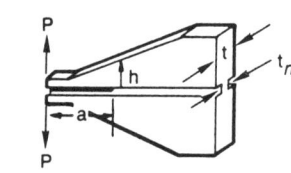

$$K_I = 2P\left(\frac{m}{tt_\eta}\right) \qquad M = \frac{1}{t} + \frac{3a^2}{h^3}$$

Figure 23. Techniques for experimental determination of mode I stress intensity factor.

TABLE 23. APPLICABILITY OF SLOW CRACK GROWTH TEST METHODS

Data Required	Critical Stress Intensity Factor, K_{IC}					Crack Growth Rate vs. K_I or $\dot{W}K_I$	
	Slow Loading Rate				Fast Loading Rate	Ambient Temperature	High Temperature
	Ambient Temperature		High Temperature				
	Porous Material	Non-Porous Material	Porous Material	Non-Porous Material			
Specimen Configuration	Tapered Cantilever Beam	Double Cantilever Beam	Double Torsion	Three (Four) Point Bend	Three Point Bend	Tapered Cantilever Beam	Double Torsion
	Double Torsion	Three (Four) Point Bend		Double Torsion		Double Torsion	
	Constant Moment	Compact Tension				Constant Moment	
		Double Torsion				Double Cantilever Beam	

Design-Related Material Technologies 309

As previously mentioned, slow crack growth can also occur during cyclic loading. To assess the damage, a standard flexure test is typically used and thermal or mechanical fluctuations are enacted to induce cyclic stresses. After a predetermined number of cycles are completed, the specimen is rapidly loaded mainly in three- or four-point bend and variations in strength are monitored. Slow crack growth damage results in a decrease in the fracture stress in comparison with a noncycled statistical strength. Cyclic loading does not always lead to crack extension and failure. Occasionally, crack tip blunting occurs during loading cycles and the result is an increase in strength.

Corrosion Test Methods

In some cases, failure of a material at elevated temperatures results from flaws or pits which were not inherent or pre-existing, but had developed during exposure to the surrounding environment. These surface pits, which may develop from preferential oxidation or attack by a chemically corrosive agent, are not functions of stressing state as is slow crack growth and may in fact occur during a period of zero stress. As the pits become larger with time, they may become the controlling flaw population and lead to failure at stresses lower than those predicted from pre-existing flaws. Such failures result from flaws which initially did not exist. Therefore, lifetime predictions based on slow crack growth of pre-existing flaws would be invalid.

To determine the time-dependence of strength in situations where the exact mechanism of failure is not obvious, a technique known as stress rupture testing is typically used. Stress rupture testing is conducted by applying a constant tensile or flexure load to the sample at a constant temperature in the environment of potential application. The time required for the sample to fail is then noted. The process is reiterated at a new stress and a different time-to-failure is observed. The stress versus time-to-failure data are then plotted at a given temperature and environment. By systematically varying the temperature and environment, a series of stress/time-to-failure plots is obtained. A drawback to the test is that the procedure involves testing a large number of specimens to obtain a complete, accurate, lifetime prediction curve covering the temperature range and environments of importance.

To reduce the number of tests, a stepped temperature stress-rupture test is frequently used to identify and screen structural ceramics for their susceptibility to static fatigue. The test is run similar to the conventional stress rupture test; however, the temperature is increased incrementally after a predesignated number of hours of dead weight loading at temperature. If the sample fails at any time during the loading cycle, its time, temperature, and stress are noted and conventional stress rupture testing is conducted under similar conditions. If the sample survives for the prescribed time period, the temperature is then incrementally increased.

The advantage of stress rupture and stepped temperature stress rupture testing is that no prior knowledge of the failure mechanism is needed to develop lifetime prediction parameters for a given material. The tests, quite simply, provide lifetime information. However, to conduct a more intensive basic research effort to answer questions about why a material failed, stress rupture testing should be used in conjunction with fractography and, when warranted, fracture mechanic type testing.

Conclusions

Test methods exist for obtaining mechanical property data on ceramic materials under laboratory conditions. However, many testing procedures require an alteration of the ceramic component to the geometry necessary for the specific test. This often alters mechanical behavior, which invalidates the test for practical application. Since current heat exchanger designs involve some tubular form, tests that adhere closely to the tubular shape should be developed. "C" ring tests offer a geometry which will not only provide practical design data, but also provide a shape which lends itself to more fundamental research (fracture mechanics, etc.).

The methods for testing ceramic materials for stress corrosion effects are not as well developed as they are for metals. Test methods need to be developed that will allow the insertion of relatively small, perhaps prestressed and/or precracked specimens, into high-temperature industrial waste streams in order to quantify the effects of stress corrosion. Further, fundamental corrosion-flaw behavior correlations need to be made on a very basic level.

Lifetime Prediction Techniques

Predicting the lifetime of high temperature structural components in ceramic heat exchangers is an important and complex feature of the design process. Exposure to extreme temperatures (>2000°F) and corrosive industrial flue gas environments can stress a material to its property limits. Many times under these environments, a structural material will undergo time-dependent degradation of strength, leading to premature time delayed failure of the component. Mechanisms of delayed failure include slow crack growth, cavitation/creep, and surface corrosion. It becomes necessary then to incorporate the time-dependence of these mechanisms, when they do exist, into the designer's evaluation of the material's applicability.

The remainder of this section will be devoted to a discussion of methods used in determining the lifetime of a structural ceramic. These include:

1. Fracture mechanics approaches, and

2. Stress rupture testing.

Fracture Mechanics Approaches

Fracture mechanics techniques have been successfully applied to electronic substrates,[50] for space craft,[51] optical fibers,[52] and ceramics for electronic capacitors.[53] Two pieces of information are necessary for using this approach: the slow crack growth behavior of the material under operating conditions and the initial flaw size present. The basic relationships used to derive the time-to-failure (TTF) involve the crack velocity-stress intensity expression and the fracture mechanics approach to strength. The equations can be defined as follows:

$$\frac{da}{dt} = AK_I^N \tag{1}$$

where A and N are constants descriptive of the slow crack growth resistance of the material and

$$K_I = Y\sigma_a a^{1/2} \tag{2}$$

where Y is a geometric constant and a is the crack size. Substituting Equation 2 into Equation 1 yields an expression for the time-to-failure.[54]

$$t_f = \frac{2}{A(N-2)\,\sigma_a^2\,Y^2\,K_{Ii}\,(N-2)} \tag{3}$$

where K_{Ii} is the initial stress intensity factor and t_f is the TTF. It is apparent from Equation 3 that the parameters can all be obtained in a straightforward manner except for K_{Ii} which requires a determination of the initial flaw size, a considerably difficult task. Two experimental approaches can be used to obtain the initial flaw size: a statistical analysis and proof testing.

Statistical Analysis. The strength of a brittle material has been shown to vary statistically. The probability of failure is dependent on the existence of Griffith flaws capable of initiating fracture. The probability of a flaw existing varies with the volume of material undergoing stress. Therefore, a close tie exists between strength and sample size. The larger the volume of material under stress, the greater the probability of a critical flaw existing.

Weibull[55,56] was one of the first to statistically analyze strength. He developed what is commonly known as the weakest-link theory of failure. Applied to ceramics, this implies that failure will occur when the strength of the weakest crack, loaded normal to its plane, is exceeded. The cumulative failure probability (P) and strength (σ_{I_c}) relationship was defined by Weibull as follows:

$$P = 1 - \exp - \left(\frac{\sigma_{Ic} - \sigma_u}{\sigma_o} \right)^M \quad (4)$$

where

σ_u = the stresshold stress, below which (P) is zero σ_u is usually equated to zero to avoid overestimations.

σ_o = the characteristic strength of a flawless sample, considered a scaling parameter.

σ_{Ic} = the inert strength of the material

m = the Weibull modulus. It is a measure of uniformity and consistency of the flaw population and is determined from the slope of the Weibull distribution plot.

Plotting the $\ln \ln [1/(1-P)]$ versus the $\ln \sigma_{Ic}$ shows the statistical scatter in strength for a given material and testing condition. A typical plot is shown in Figure 24.

Design-Related Material Technologies 315

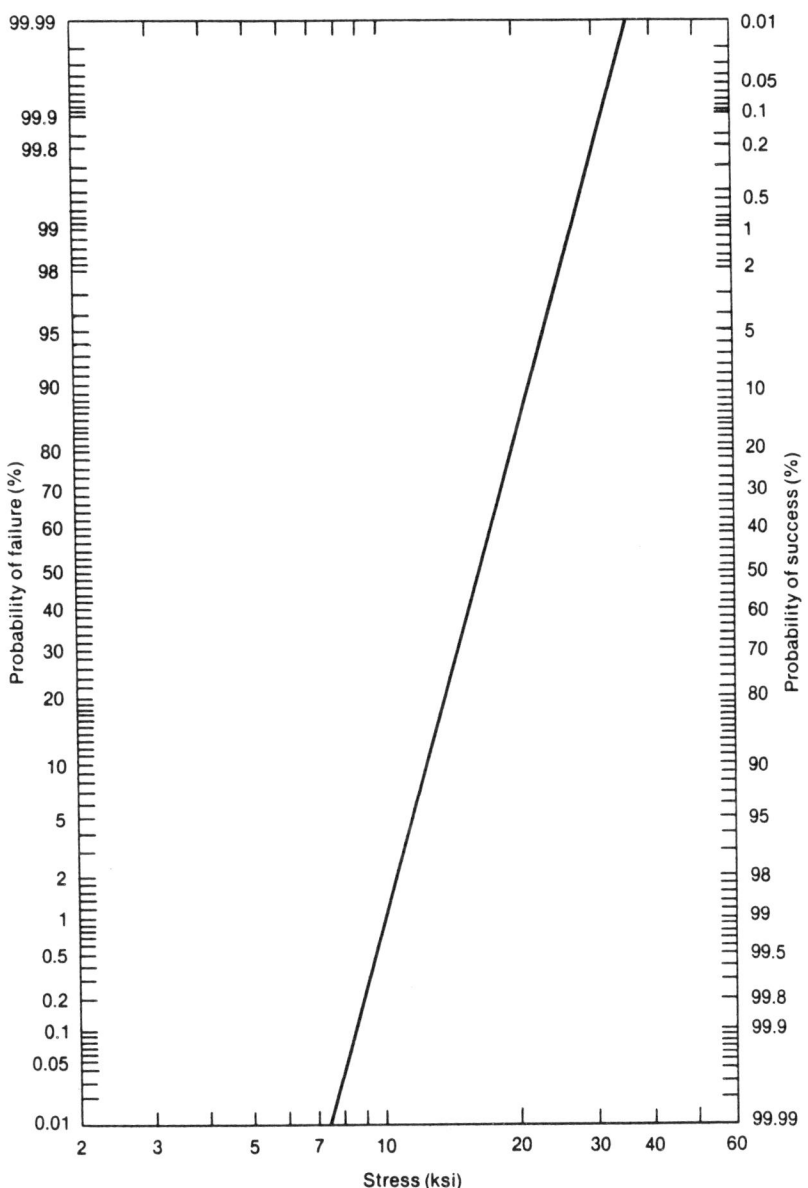

Figure 24. A typical Weibull plot.

Lifetime predictions are made by expressing K_{I_i} in terms of Weibull-generated failure probability.[57,58] The following relationship is generally used at failure:

$$K_{Ic} = \sigma_{Ic} Y a_i^{1/2} \tag{5}$$

where K_{Ic} is the critical stress intensity factor. Solving for $a_i^{1/2}$ and substituting into Equation 2 yields

$$K_{Ii} = K_{Ic} \frac{\sigma_a}{\sigma_{Ic}} \tag{6}$$

By substituting Equation 4, with $\sigma_u = 0$, into Equation 6 an equation relating K_{I_i} to the failure probability is obtained:

$$K_{Ii} = K_{Ic} \frac{\sigma_a}{\sigma_o} \log \frac{1}{1-P}^{-1/m} \tag{7}$$

The time-to-failure is then related to the cumulative probability of failure:

$$\log t = \frac{N-2}{M} \log P - N \log \sigma_a + \log \Phi \tag{8}$$

where

Φ = a constant based on material parameters

P = the probability of failure

A more descriptive form of this equation as expressed by Wiederhorn[59] is:

$$t = \sigma_a^{-N} f(P) \tag{9}$$

Plotting log t versus log σ_a at different values of probability yields a series of lines of slope (-N) and describes the failure characteristics of a material on a statistical basis. A typical plot is shown in Figure 25 illustrating the applicability of the technique (Reference 58). It is very useful for design purposes since the entire range of strengths is represented rather than just an average or median value.

In applying Weibull statistics to failure predictions several precautions must be used. The parameters M and σ_o are important in determining the strength estimates; therefore, they must be highly accurate. Several potential problems exist in the use of m and σ_o. These parameters may vary from one component to the next due to differences in manufacturing processes. It may therefore be inaccurate to extend M and σ_o values determined on small lab samples to actual material components. Flaw populations must be well known in order to apply scaling equations which are often very complex. If failures occur from edge flaws, serious errors in failure probability estimates will result. Multiaxial stress states present another source of problems. Under these stress conditions, simple Weibull theory, which assumes tensile failure of the most severe flaw, may not apply. Failure may result from smaller flaws more favorably oriented in the stress field. This behavior has been addressed recently by Batdorf[60] and Sines.[61] These problems and others generally lead designers to use more conservative estimates when statistical approaches toward lifetime predictions are used.

Weibull statistical methods have been applied to ceramic heat exchanger design. Garrett AiResearch has used Weibull statistics to the greatest degree for heat exchanger design proof of concept. Garrett uses a finite element model (ANSYS) to calculate a thermal and mechanical stress

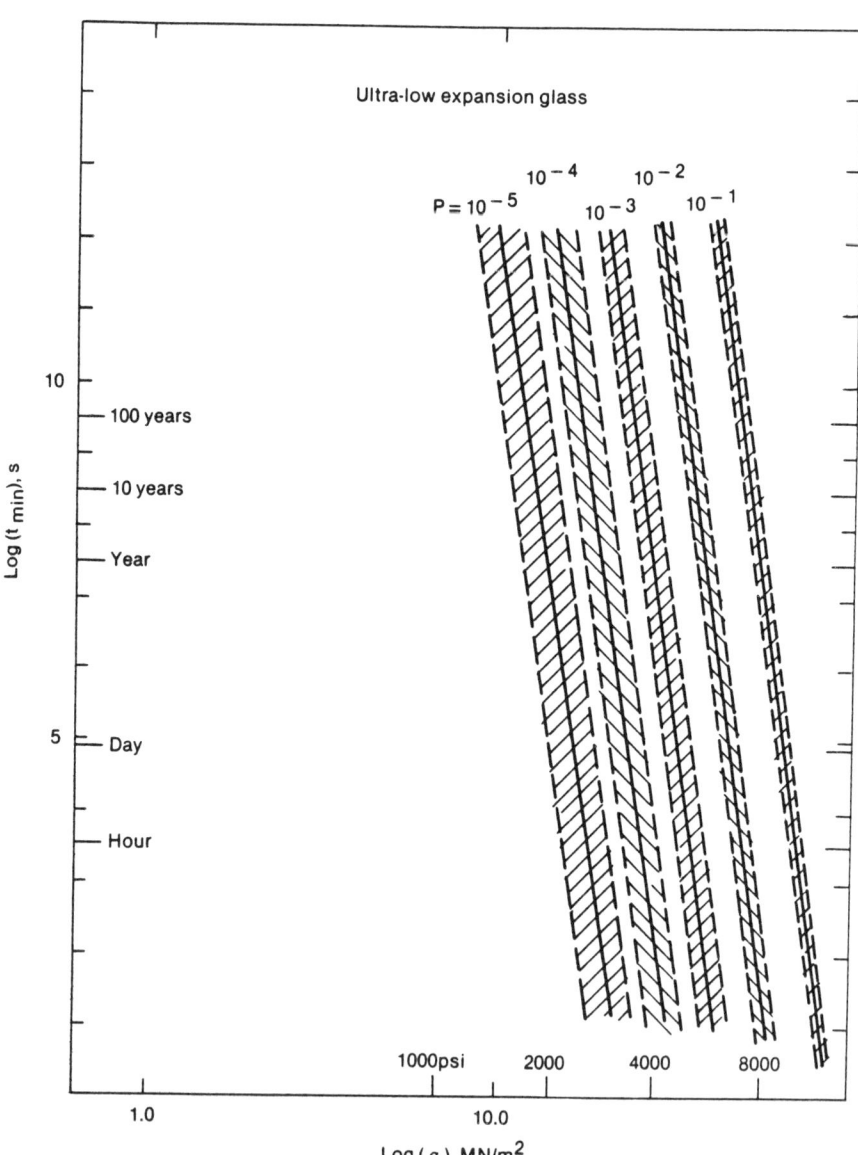

Figure 25. Statistically based design diagram. For a given value of the failure probability, each line on this diagram relates the failure time to the applied stress; the cross-hatched area gives 95 percent confidence limits for the position of each line.

distribution map for each high temperature component. A proprietary Garrett computer program (ANSTAT) calculates a probability of failure for each element using strength data taken from Garrett's extensive data base. The data base consists basically of three- and four-point Modulus of Rupture data gathered by Garrett. The computer program then sums up all the probabilities over all the elements and calculates a probability of failure for the entire component. Lifetime prediction estimations are made by calculating probabilities of failure using different loading rates. This will reveal a trend showing increased probability of failure with slower crosshead speeds. These trends are then extrapolated to give probabilities of failure of an anticipated lifetime.

Solar Turbines has used Weibull statistics to compare strengths of Carborundum's KT silicon carbide in a radial compression test at room temperature. They used compliant interlayers to prevent point loading between the ceramic tube specimens and the testing fixture. When the compliant interlayers were not used, the Weibull modulus was reduced. They also noted that the Weibull modulus was significantly increased when fast fracture radial compression tests were performed at 1200°C.

Solar Turbines also used Weibull statistical analysis to test Norton's NC-430 in a radial compression test at room temperature and at 1200°C. They found, by comparing Carborundum's material to Norton's material, that probable failure loads with temperature, at 0.5% failure rate, indicates that KT silicon carbide is stronger than NC-430.

Solar Turbines concluded that it is not possible to directly apply the failure probability data from the radial compression samples to determine probability of failure of tubes 4.6 m (15 ft) long. The importance of these tests was the ability to compare relative strengths of selected ceramic materials.

Hague International, Babcock & Wilcox, and GTE have done very little work in the area of Weibull statistics. Babcock & Wilcox has used data generated by other researchers to compare materials for selection. It should be noted that since the material chosen by GTE (cordierite) has a very small coefficient of thermal expansion, thermally induced stresses were deemed insignificant.

Proof Testing. A more absolute technique of determining the initial flaw size and thus K_{Ii} is proof testing (Reference 59). The use of state-of-the-art proof testing techniques to improve mechanical reliability is well known[57,59,62-66] and is accomplished by removing weak components from the final population by the proof test. This action modifies the original strength distribution by truncating the lower end through removing weak components from the population before they are put into service. This technology is proven and has been successfully applied to a number of widely varying materials and applications, from the windows of the space shuttle (Reference 51) to electrical insulators.[67]

The application of proof testing to ceramics components was initially by trial-and-error (Reference 59). This empirical approach to determining acceptable proof stress levels was adequate in that it provided reliable post-proof test life, but it lacked any true systematic basis which could be used by advancing technologies. The efforts of Wiederhorn (Reference 59), Evans (References 54 and 57), and Ritter[68] provided a mathematical foundation for the application of proof testing to ceramic materials, both in selection of proof stress levels and proof test conditions. A natural extension of this quantitative approach to proof testing is the application to design.

In proof testing a specific component, several decisions must be made with respect to the manner in which the test will be conducted, i.e., atmosphere, loading rate, hold time, unloading rate, and maximum proof stress.

The following list (Reference 59) summarizes those assumptions that one makes when applying a proof test:

1. The proof test must duplicate the actual stresses expected in the component;

2. The proof stress exceeds the design or service stress by a specified amount; and

3. The failure mechanism that occurs inservice must be the same as that by which proof test failures occur.

The following approach is used to apply proof testing to determine the initial flaw size. Proof test survival guarantees that the stress intensity associated with the largest flaw did not exceed the critical values K_{Ic}, thus, Evans and Wiederhorn (Reference 57) show that

$$\frac{K_{Ii}}{\sigma_a} = \frac{(K_i)_{proof}}{\sigma_p} \quad \frac{K_{IC}}{\sigma_p} \tag{10}$$

where

σ_a = applied stress

σ_p = proof stress

K_{Ii} = initial stress intensity

K_I = maximum stress intensity during proof test.

Substitution of $K_{Ic} \frac{\sigma_a K_{Ic}}{\sigma_p}$ into Equation 3

for K_{Ii} produces an expression for the minimum time to failure

$$t_{min} = \frac{2 \sigma_p/\sigma_a{}^{n-2}}{(n-2)A\sigma_a{}^2 Y^2 K_{Ic}{}^{n-2}} \tag{11}$$

The dependence of minimum life on the ratio of proof stress to applied stress allows the graphical representation of minimum time to failure as shown in Figure 26 (Reference 57). This diagram, developed for soda-lime glass in water, allows the maximum proof stress to be selected from the design parameters of minimum life and applied stress. Further extension of the TTF described by Evans (Reference 57) include analytical methods to determine strength after proof testing, time to failure without proof testing, and time to failure after proof testing (for TTF > Minmimum TTF).

A major problem with proof testing is that it characterizes only the initial flaw population. Material that undergoes changes in flaw populations when exposed to high temperatures and/or severe environments do not follow predicted failure probability and therefore lifetimes cannot be accurately determined. For example, Tighe and Wiederhorn[69] have found the proof testing technique to be valid for hot-pressed and reaction-banded Si_3N_4 at room temperature. However, a high temperature (1200°C) exposure resulted in a flaw population that was independent of that observed in the as-machined samples. Therefore, the reliability of proof testing was diminished. Another difficulty arises when slow crack growth occurs in the proof stressed material especially during the unloading of the proof stress. Slow crack growth during loading and holding times is not as critical since the flaw will still be smaller than the critical size set by the proof stress. However, slow crack growth during unloading can result in a flaw size, close to the critical size, extending beyond critical and therefore damaging the reliability of the test. Evans and Fuller (Reference 54) have discussed these conditions in detail.

Proof Testing: Applications. Proof testing is a viable method of determining the life of a critical component. This is evidenced by previous applications. The following examples are for applications of ceramic materials whose failure could result in either injury, loss of life, or large capital losses.

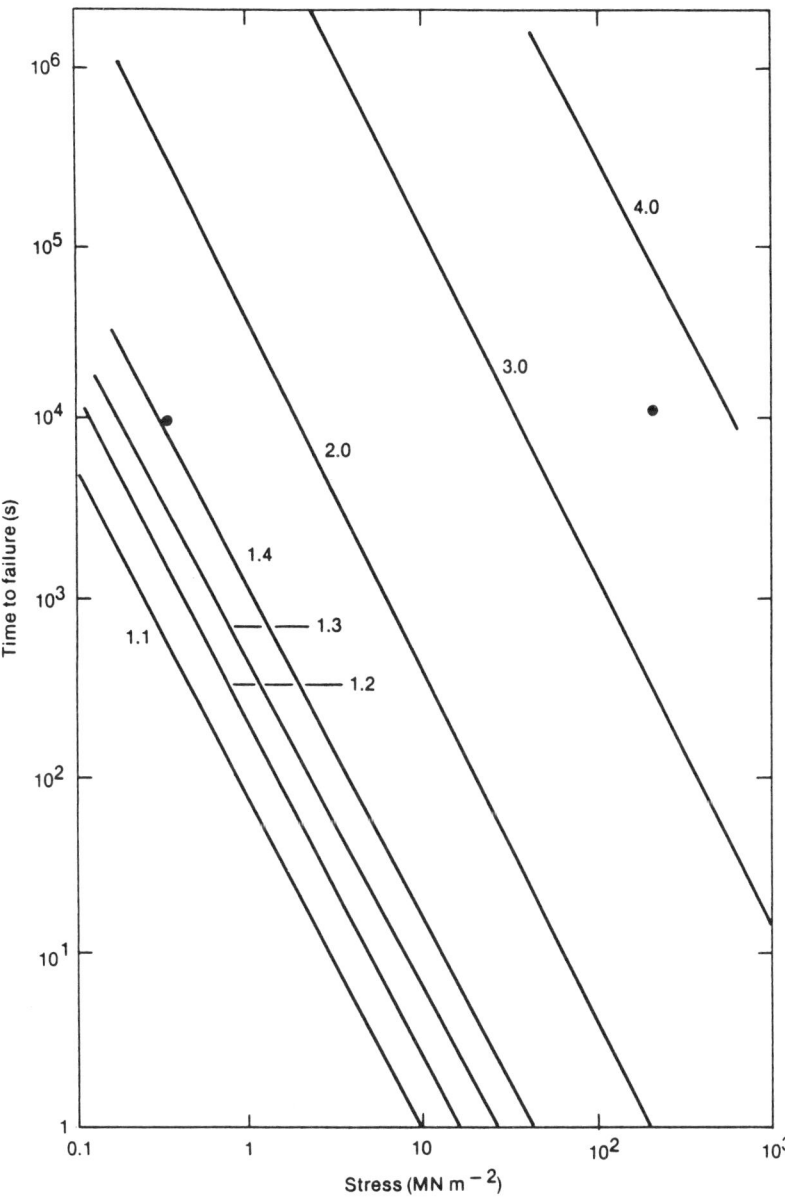

Figure 26. A proof test diagram for soda-lime glass in water. The numbers below each line refer to the proof stress/applied stress ratio (σ_r/σ_s) corresponding to that line.

One particularly significant application of proof testing was the evaluation of windows for the space shuttle (Reference 51). A minimum lifetime of 1 year in 100% relative humidity was assured when a proof test was employed to a proof stress of ~3 times the design stress. Long-term predictions of lifetime required extrapolation of crack velocity data beyond experimental limits. Thus, the functional dependence of stress intensity on crack velocity was determined with adequate reliability such that extrapolations did not result in excessive errors in lifetime prediction.

Ritter and Wulf (Reference 66) confirmed, in a study of vitreous-bonded abrasives, behavior consistent with predictions from fracture mechanics theory. Their results indicated failure by slow crack growth of preexisting flaws. During this study it was shown that proof stress ratios which were currently in use were adequate to assure the design lifetimes. In other studies, critical ceramic components were successfully proof tested for gas turbine applications.[70] The minimum strength was guaranteed in 150-km lengths of optical fibers by a proof testing program conducted in Japan.[71]

Wynn, et al. (Reference 70) point out that proof testing is the most reliable method of eliminating unacceptable components if certain conditions are met. These conditions are:

1. The proof test must generate loading such that the proof test failure mode is identical to service proof test modes;

2. The proof test must reject faulty parts with certainty, thus sacrificing some good parts;

3. The proof test must not adversely affect the subsequent service life of the component;

4. The proof test must be reasonable from a cost standpoint; and

5. Proof test survivors should have an increased probability of survival in service as a population.

The presentation of these conditions is considered illustrative in that this list includes many of the assumptions that are required when performing a proof test.

Other applications of proof testing include the evaluation of high temperature exposure in air of hot-pressed silicon nitride.[72] A potential maximum proof test stress has been recommended for hot-pressed silicon nitride as a result of other proof testing work (Reference 69). The effect of prestressing of siliconized silicon carbide was also evaluated via proof testing.[73]

Stress-Rupture Testing

Stress-rupture testing is a useful technique for determining time-dependent properties of brittle materials.[74,75] The technique is a straightforward approach in which few assumptions are made and a particular mechanism of time-dependent failure such as slow crack growth does not need to be assumed.

Stress-rupture measurements are made by monitoring the time-to-failure of a specimen under a constant load, in the temperature and environment of interest. By changing the constant load from test to test, a range of time-to-failure values is obtained, provided a time degradation of strength is occurring. A plot of time-to-failure versus applied stress is generated as shown in Figure 27. A least-squares fitted line through the data points generally infers the following relationship:

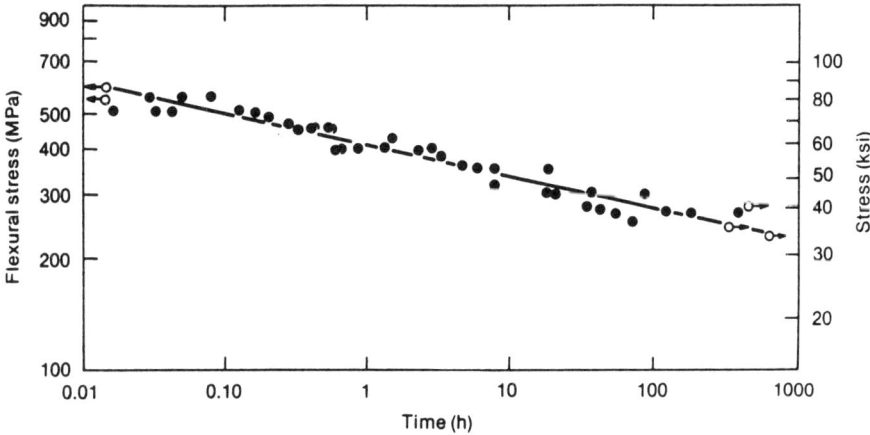

Figure 27. Flexural stress rupture at 1200°C for Norton NC 132 hot-pressed silicon nitride.

$$t_f = B\sigma_a^{-N} \qquad (12)$$

where

t_f = the time-to-failure

B = a constant

σ_a = the applied stress

N = the stress intensity exponent from Equation 1.

The time to failure can therefore be predicted for a given applied stress using the relations in Equation 12 regardless of what mechanism of delayed failure is operative.

Mechanisms of failure or degradation can be identified by performing detailed fractography on the fracture surface.[76] Sources of failure such as pits, pores, or inclusions can be determined and crack growth mechanisms such as slow crack growth can be identified.

The interpretation of stress-rupture data requires some precautions. A material undergoing considerable permanent deformation will, in essence, reduce the maximum tensile stress on the sample. The predicted time to failure may therefore be slightly higher than actual values. Some safety factor should be included, particularly for long time situations. A second complication results from assumptions made that the crack shape and stressing state do not change throughout the course of the test. These assumptions can also lead to some variations between the predicted and actual time to failure.

Conclusions

Fracture mechanics approaches and stress rupture testing are both viable techniques of making lifetime predictions based on time-dependent strength degradation of a material. Both techniques have advantages and disadvantages pertaining to their applicability as were discussed previously.

The lifetime prediction technique that appears most suitable to the design of ceramic heat exchangers is stress-rupture testing in conjunction with detailed fractography. Strength degradation and lifetime predictions can be determined directly from the tests and microstructural analyses of ceramic specimens that are exposed to temperatures and/or environments of importance. There is no need to make preliminary assumptions on mechanisms of failure as is necessary when using fracture mechanics. There is also no need to worry about changes in flaw populations during the test. The results from stress-rupture testing can therefore be incorporated conveniently into design methodology.

Product Reliability Techniques

The ability for the designer to be assured that a purchased ceramic component will exhibit a given durability within standard specifications is critical. Because of the wide statistical variability in properties of ceramics, it is especially necessary to have a production scale technique for eliminating the worst ceramic pieces, thus assuring greater durability by narrowing the range of mechanical and lifetime properties that a designer can expect. This would also reduce the degree of necessary design conservatism.

The following section assesses the ability of two types of currently available product reliability techniques--nondestructive evaluative techniques and proof testing--to provide the designer with greater assurance of a ceramic's mechanical properties.

NDE

The brittleness and statistical variation in fracture toughness for ceramics makes a cost effective NDE inspection highly desirable for production-scale preservice and inservice inspection of ceramic heat exchanger tubes. An effective NDE technique will ensure that heat exchanger tubes, for instance, will meet a minimum performance specification. As with any step in the manufacturing and use process, it must be cost effective. The value or cost of NDE can be evaluated both with respect to pure dollar savings from failure prevention and with respect to factors such as promoting industry acceptance or proving the capability of a device. This section examines these questions to provide some basis for determining what level of NDE effort is appropriate for the problems associated with heat exchanger quality assurance.

Physics of Ceramics NDE. The primary characteristics of ceramic materials that are relevant to the application of nondestructible evaluation methods can be summarized as follows:

1. Typically, ceramics have a small critical flaw size. Depending on the design and stress environment, the critical flaw size can be 10 μm or smaller. (However, for heat exchanger tubes in a low stress environment the critical size may be much larger; estimates range from 100 to 1500 μm, which is still small compared to metals.)

2. The sonic velocity of ceramics is approximately twice that of metals. This affects the application of many acoustic techniques for two reasons. The wave length of sound increases with velocity, reducing sensitivity to small reflectors, and the refraction of sound at interfaces is increased, resulting in potential increased signal loss at angular or rough interfaces.

3. Structural ceramics are, in general, nonmagnetic and nonconductive, which limits the utility of induced electromagnetic methods.

4. Some ceramic fabrication methods such as slip casting can result in a nonsmooth surface which can reduce the accuracy of x-ray and ultrasonic methods.

5. Ceramics can be porous, which reduces the efficiency of techniques such as dye-penetrant.

Currently Applied Technology in Production. NDE techniques currently used in the production inspection of ceramics are direct adaptations of techniques used on more conventional materials. X-ray radiography, dye penetrant, visual and microscopic inspection, and leak testing are generally the only techniques used regularly for production quality assurance of ceramics. There are other useful techniques being applied on a smaller scale for specialized purposes, but these are not generally applicable to ceramic inspection. Several references[77,78] give an overview of the various techniques that have some application to ceramics, and the reader is referred to these for more detail.

For heat exchanger tubes, such as those manufactured by Norton and Carborundum, the only NDE techniques currently applied are dimensional measurement, visual and penetrant examination, and conventional radiography. None of these are capable of detecting anything but relatively gross defects. The users such as B&W, Garrett, and heat exchanger fabricators have no additional capabilities and in fact do no substantial additional NDE.

Technology Under Development. The NDE methods mentioned previously are inadequate in the sense that there is an almost universal desire on the part of ceramics manufacturers and users for better NDE of their products. While many of these techniques are adequate for detecting gross flaws it is either impossible or tedious to use them for the small critical flaws typical in ceramics. It is significant that ultrasonics has not been applied to production inspection of ceramics as it is one of the most widely used methods for more conventional materials. This is primarily due to the higher sonic velocities required and small critical flaw sizes discussed earlier. However, much of the current development work is based on acoustic methods.

An assessment of the developing technologies reveals that there is no organization currently capable of providing a general NDE capability for structural ceramic components that could be utilized by industry. While many of the elements for an effective NDE system exist, they are not being integrated in a way to take maximum advantage of their potential. This is understandable in that many industrial laboratories need to focus on specific problems while many of the national laboratories are geared towards more basic research. Nonetheless it is possible to identify the most promising techniques and to suggest an approach to provide maximum industrial impact. An effort is needed to bridge the gap between laboratory research and the real inspection needs of ceramic component manufacturers. The technology is developed enough to apply "engineering research" with the objective of providing prototype inspection capabilities. If these advanced techniques can be evaluated by manufacturers, a great deal of valuable information can be obtained to begin the technology transfer. A brief review of methods under development and their status follows.

Ultrasonic Scanning--A number of government and industry laboratories are working on the development of NDE techniques for ceramics. It appears that the efforts can be divided into projects that are either very basic in nature or overly specific and limited to a particular problem. One notable exception to this division was the work performed by Kupperman at Argonne East and Yuhas at Sonoscan for the Coal Conversion and Utilization Technology program. Through fiscal year 1982, this work attempted to survey and develop applied NDE techniques for ceramic heat exchanger tube geometries. The most notable hardware development was a bore-side ultrasonic scanning device that operated at 20 MHz. While the scanner was capable of detecting flaws on the order of several hundred microns, it was very sensitive to variations in wall thickness geometry. This research group has recently shifted their emphasis to an examination of the more fundamental NDE problems in ceramics such as the characterization of "green state" materials. While they have provided a wealth of important information on the particular techniques they investigated, it was neither the purpose nor result of their work to provide a prototype capability to industry that could be immediately applied to ceramics inspection.

Scanning Photo-Acoustic Microscopy--Industrial NDE research being conducted is generally geared towards specific problems arising from the fabrication of turbine blades and related parts from ceramic materials. Scanning photo-acoustic microscopy (SPAM) and higher frequency conventional ultrasonics have been used to give some support for product development. In general, though, there is no generic capability being developed due to the very difficult nature of the problem. Kandeleval and Heitman of Detroit Diesel Allison (DDA) and Silversmith and Wakefield of Gilford Instrument Laboratories[79] describe the application of SPAM to structural ceramics in laboratory tests. Currently, DDA is continuing development of this technique along with high frequency acoustic imaging, but there remains considerable work to be done before the results can be generally applied. This work was supported by NASA through the Ceramic Advanced Turbine Engine (CATE) program.

NASA has also performed their own assessment of ceramic NDE technology[80] and is conducting basic research in the use of acoustic and microfocus x-ray technologies applied to ceramics. In general, however, this work is limited to laboratory samples and is more in support of materials research rather than the development of a generally applicable NDE Technology.

Pulse-Echo Acoustic Microscopy--In related work, pulse-echo acoustic microscopy has been applied at the Idaho National Engineering Laboratory to production inspection for very small flaws (50 µm) in nuclear fuel plates. The acoustic system is similar to those being used at NASA Lewis, DDA, and GE Research Labs for NDE development for ceramics. The differences are in the unique "fast pulse" signal processing and system design that allows for daily use in a production environment. While this system has seen limited application to ceramics per se, it does appear to have solved many of the problems involved in applying the "delicate" measurements in a production system to provide a generally applicable NDE capability.

R. Gilmore at GE Research Labs has been developing a similar pulse-echo acoustic microscopy system for use in a wide variety of materials. (It is interesting to note that NASA Lewis, DDA, INEL, and GE are all approaching the ceramics inspection problem with the same or similar acoustic transducers. In fact they are all from the same supplier.) The GE system is still essentially a laboratory tool although they are developing a broad base of applications, some of which will be in the field. Depending on the application and material, GE's acoustic microscope may actually utilize fairly low frequency ultrasonic signals to provide a gray scale c-scan of the interior of components. Their published results, however, have not specifically addressed the problems of refraction and surface roughness associated with ceramic heat exchanger geometries and they have not gone beyond standard ultrasonic instrumentation techniques for their signal processing.

Microfocus X-Ray--Microfocus x-ray differs from conventional x-ray radiography primarily in the smaller spot size of the x-ray generation tube. While conventional x-ray systems typically have a 0.6 to 5.0-mm spot size, a microfocus head may have a spot size as low as 10 μ.[81] The spot size is the diameter of the area on the x-ray tube target where the x-rays are generated by bombardment with a focused beam of high energy electrons. The smaller size results in greatly improved resolution in the image. It is possible to image defects as small as the spot size if the defect has an x-ray attenuation sufficiently different from the surrounding material.

The small spot size of microfocus systems limits the flux and energy level of the x-ray beam generated due to the need to maintain the temperature of the target below the melting point. The smaller target area concentrates the heat generated for a given incident electron beam flux. Because of this, the penetrating power of microfocus systems is limited but is still within the range to be useful for ceramic heat exchanger tubes with a thickness on the order of a centimeter or less. However, recent advances in tube-head design are yielding microfocus systems capable of operating up to 160 kV potential which will undoubtedly widen their applicability.

The image formation scheme with a microfocus x-ray system can be normal fine grained x-ray film or an x-ray sensitive television tube such as a Vidicon. The television scheme permits "in motion" examination of components, which in many cases improves flaw detectability. In addition, the closed circuit TV image provides a magnification which also aids in detectability. The optimum resolution, however, is obtained with film.

Acoustic Microscopy[82]--Acoustic microscopy can be classified under one of four techniques. All of these represent the growing effort to apply acoustic techniques to ceramics NDE.

Scanning Laser Acoustic Microscopy (SLAM) is perhaps the most well developed of the acoustic methods. It is primarily championed by its developer, Sonoscan, of Bensenville, Illinois. In this scheme a sample to be imaged is prepared and placed on the microscope stage. If the sample surface is not sufficiently light reflective a reflective cover slip is coupled to the sample. High frequency ultrasound (usually about 100 MHz) generated by a transducer is then sent into the part. An optical scanning system sweeps a laser beam over the coverslip and the reflection is detected and converted to a TV image. The image formation represents the mechanical vibration of the part surface caused by the ultrasound beam propagating through the part. Any defects in the part being inspected represent a change in acoustic impedance which disturbs the sound propagation through the part. If the defect is large enough and close enough to the surface being scanned, the surface vibrations are perturbed or shadowed and the variation can be detected in the image.

The advantage of the SLAM is in its ability to detect flaws as small as 10 μ and its appealing real-time "image." However, it must be kept in mind that the image is really a map that corresponds to surface vibrations and is subject to interpretation particularly if a flaw is deep in the part. The disadvantages of the technique are primarily that the sample or part being inspected usually must conform to the inspection system size and configuration limitations. Kupperman and Yuhas have investigated the design of a production configuration for SLAM of ceramic heat exchanger tubes. While it is technically feasible to design equipment to inspect large heat exchanger tubes, it would be a significant engineering task. Nonetheless it is a well developed and useful system for detecting and characterizing small flaws in appropriate situations.

In a Scanning Photo-Acoustic Microscope (SPAM) the roles played by laser and acoustic transducer are essentially reversed from that found in the SLAM. In this case, the laser beam is slowly scanned across the part being inspected and a low frequency acoustic transducer is used to detect stress waves caused by intense localized heating of the part surface where the laser beam is incident. The scanning is accomplished by translating either the part or the laser. The signal received by the transducer is used to form an image line by line as the scan progresses. With this method, image formation is slower than with the SLAM although the resolution is comparable. The phenomenon is sensitive only to surface or near-surface defects that are within the locally heated region. It would be difficult if not impossible to adapt this technique to a rapid inspection of full size heat exchanger tubes, especially with its limitation to surface defects.

In Pulse-Echo Acoustic Microscopy (PEAM), a single high-frequency ($\geq$50 MHz), sharply-focused ultrasonic transducer is used both to inject a pulse of high frequency sound into a part and to receive reflections of the ultrasound from any material discontinuities. Image formation or flaw detection is accomplished by scanning the part or the transducer and recording or otherwise analyzing the information in the echos. PEAM is capable of better resolution than SLAM for a given ultrasonic frequency. The technique is amenable to high speed production volumetric inspection even though its image formation is not as rapid as with a SLAM. Since a single transducer is used to send and receive the interrogating energy, the scheme is inherently less complex than either SLAM or SPAM. PEAM does require that the inspection take place in an acoustic coupling medium such as water, but this would not preclude its use for ceramics. The resolution for imbedded flaws is ultimately determined by the acoustic wavelength in the part being inspected. For example a 100-MHz center frequency transducer would have a sonic wavelength of 100 μ in silicon carbide (NC340) and the best expected resolution might be one-half of the wavelength. A 200-MHz transducer would do twice as well. Resolution for defects open to the surface, however, is determined by the wavelength in the coupling media, which generally results in a wavelength about seven times smaller than that in the ceramic. Detection of an imbedded defect that is close but not open to the surface would be limited by the duration of the ultrasonic pulse, a phenomenon known as near surface resolution limitation. This is due to the fact that no returning echoes can be detected until the reflection from the surface of the ceramic has died down and left the signal detection circuits quiescent. The depth of this "blind spot" is dependent on the transducer bandwidth and material velocity, but would typically be on the order of 25 to 200 μ into the ceramic.

In Transmission Acoustic Microscopy (TAM) two ultrasonic transducers are used in a transmit/receive mode. In general, this requires access to two opposite sides of the sample which is immersed in a coupling medium such as water. Essentially it functions similarly to PEAM except that the information being analyzed is derived from the energy transmitted through the part rather than reflected energy. While it would undoubtedly be more sensitive to certain types of flaws such as angular reflectors or cracks, it is more difficult to apply to production inspections due to the two-transducer configuration.

Conventional Ultrasonics--There has been development work undertaken for applying more conventional ultrasonic techniques to ceramics at medium range frequencies on the order of 10-20 MHz.[83,84,85] Some of this has been specifically geared to heat exchanger geometries with specific hardware and scanning devices. While they have demonstrated the capability to detect flaws on the order of a 100 μ, there are serious practical limitations due to part irregularities such as tube wall thickness variations. While all of the NDE techniques discussed are sensitive to surface roughness and dimensional variations, medium frequency ultrasound may suffer more due to the small signal to noise ratio for tiny reflectors. It is not anticipated that medium frequency techniques will be useful except for the detection for relatively large defects or for less complex problems such as determining part thickness.

Acoustic Emission (AE)--AE has not been widely applied to NDE flaw detection in ceramics, apart from material testing, since in general it requires actual crack growth or dislocation to occur before it can be used. Essentially acoustic emission methods rely on the detection of high frequency sound generation by energy release in the material being inspected.[86,87] It can be highly useful however during proof testing to determine if any flaw growth or formation has occurred. Moreover, continued research in this area may result in the ability to detect small energy releases that are characteristic of flaws without the need for actual flaw growth.

Other NDE Methods--Techniques such as thermal imaging, dielectric measurements, and microwave, while they have been investigated, do not appear to be particularly promising for inspection of ceramic heat exchanger tubes. This is primarily due to the insensitivity of the interrogating energy to the small flaw size requiring detection and the rather specialized characteristics of the methods.

Cost Effectiveness of NDE. A very rough order of magnitude estimate for the cost of failure of a ceramic heat exchanger tube in a typical situation is presented. The purpose of this estimate is to give some idea of what can be saved by preventing failure through NDE. The lack of detailed engineering and performance data makes it difficult to be precise but it is possible to estimate the scope of the problem.

The ceramic heat exchanger tubes fabricated by Norton for use in the B&W heat exchangers cost approximately $1230 per tube on the most recent contract ($1325 for the outer tube; $1135 for the inner tube). While the per tube cost would certainly be reduced by mass production, this average figure can serve as a starting point for an estimate of what might be a reasonable amount to spend on NDE in this situation. If we assume the heat exchanger is being used in a steel mill soaking pit, a typical repair of a broken tube would take approximately 76 manhours or about $1900 plus materials.

The next step in the estimation process is much more difficult because it requires an estimate of how often a tube might break during the heat exchanger lifetime, and there are no reliable data to draw upon. In addition, an assumption must be made as to how far the failure rate could be reduced by effective NDE using estimates based on experience and engineering knowledge. For example, if a typical heat exchanger consists of 200 tubes with a design lifetime of 10 years and an average of 5% of the tubes fail per year due to a preservice defect, then 10 tubes per year will fail, or 100 over the lifetime of the exchanger. If a preservice inspection would reduce this failure rate to 1% per year, then the inspection will have saved the cost of materials and repair for 80 tubes. This simplified way of examining the problem would indicate that approximately $1230 + $1900 x 80, or $250,400 could be saved by performing a preservice inspection that would reduce the failure rate to 1% per year. If this is prorated over the 200 tubes, it would imply that the breakeven point would be at $1252 of NDE cost per tube. Examples for other failure rates are shown in the table below.

NDE COST ANALYSIS
200-TUBE HEAT EXCHANGER

Uninspected Failure Rate Percentage (tubes/year)	Failure Rate After NDE Percentage (tubes/year)	Number of Failures Avoided Over 10 Years	Savings	Breakeven NDE cost/tube
5 (10)	1 (2)	80	$250,400	$1,252
4 (8)	1 (2)	60	187,800	939
3 (6)	1 (2)	40	125,200	626

Of course it is highly speculative at this point as to what the failure rates will be and what improvement would result with an effective NDE program. Nonetheless, an analysis similar to the one above, with additional detail and refinement, should be used in evaluating the cost effectiveness of NDE methods.

Some of the NDE methods proposed for heat exchanger tubes are quite expensive in the research laboratory. This expense does not mean they should be ruled out, but the development work undertaken should incorporate methods to reduce both initial equipment and operating costs.

Another, and perhaps more important, aspect to consider is how to evaluate the intangibles such as the perception and acceptance industry has for a new technology. NDE can impact these by guaranteeing that pilot units will not suffer excessive failure which is not representative of potential reliability. It becomes much more difficult to estimate the worth of NDE under these circumstances, but such worth should not be discounted.

Proof Testing

Proof testing has been shown to be a valuable tool in lifetime prediction theory; however, it is more commonly used to ensure product reliability or quality assurance (QA). Proof testing is used to eliminate weak components prior to being placed in operation since they are the most likely to fail in service.

Many of the guidelines discussed earlier still apply and should be followed. Proof testing for QA involves applying a proof stress to each component greater than the stress the component will see in service. Component containing flaws of a certain critical size will fail. The components which survive will have a zero probability of failure at the expected in-service stress unless time-dependent failure mechanisms were operative.

The proof testing of ceramic tubes as a QA procedure would most likely involve the pressurization of each individual tube to a stress exceeding that expected in service. Tubes which failed catastrophically or were unable to maintain pressure as the result of a leak would be eliminated, leaving only those tubes more suitable for application.

Several precautions must be taken when using proof-testing for QA. The proof stress chosen must be reasonable, respective to the in-service stress. Too-large proof stresses will lead to unnecessary breakage of good tubes. The proof test must be applied such that crack growth does not occur during the unloading cycle. Crack growth during unloading can lead to flaw sizes larger than the critical size of the proof stress.

Of the many manufacturers, only GTE-Sylvania performed what would be defined as a proof test.[88] That is, the application of some stress greater than the design or operational stress to the final assembly. The procedure employed by GTE-Sylvania is performed in three steps. An initial static leak check to ~2 psi is performed to indicate overall exchanger integrity. This procedure is followed by a static burst test to 10 psi which is followed by a 2-psi dynamic leak check. The dynamic leak check serves to open heat exchanger tube flaws not normally identified during the static leak check. The maximum pressure of 10 psi is based on a combination of strength at operating temperature and the design factor of safety.

Garrett AiResearch performs a pressurization of tubes prior to assembly, the intent of which is to locate flaws.[89] Babcock and Wilcox feel that any application of a proof stress will potentially introduce flaws, and therefore do not perform any proof test. Their final assembly is only leak checked.

Assessment and Future Objectives

It is clear that reliable, cost effective NDE techniques do not currently exist to detect crucial flaws in ceramics except in a few cases. Standard radiography and visual inspection are probably the most widely used and these are either insensitive to small flaws or tedious and labor intensive to use.

There are several promising techniques under development, but none at the present time are useful for inspections outside of a controlled laboratory environment. The most promising of these are microfocus x-ray radiography and acoustic microscopy. Acoustic microscopy requires the most development but also holds the most promise for general applicability. The optimum near term benefit for ceramic heat exchanger tube NDE will most likely be derived from the application of pulse-echo acoustic microscopy techniques and microfocus x-ray radiography. PEAM has the most versatility for production inspections, while microfocus can provide a complementary, high resolution image.

A cohesive and effective NDE development program to support the commercialization of ceramic heat exchanger tubes should have as its basis the following four objectives.

1. Engineering research should be emphasized to coordinate, utilize and expand the basic research presently being conducted. The thrust of this research should be to provide a practical advanced NDE capability that can be used to begin the technology transfer to industry and to identify the major remaining problems in production application of NDE to ceramics.

2. Fabrication and NDE technology should be coordinated so that the shape and surface condition of the as-fabricated tubes are amenable to inspection when inspection is necessary. All of the NDE techniques under development are sensitive to surface effects to a much higher degree than that found in metals. This policy should be implemented in a way that does not overly complicate the design and fabrication of ceramic heat exchangers. But at the same time, the designer should have eventual NDE in mind, particularly when considering surface finish.

3. NDE and fracture mechanics should be related to develop consistent and comprehensive engineering design data. This step will provide a methodology for distinguishing critical flaws during production inspections.

4. Development should be concentrated on high frequency ultrasonic and microfocus x-ray techniques.

5. An ongoing evaluation of other techniques that are currently not practical should be continued. These include thermography, acoustic emission, and electromagnetic techniques other than x-ray radiography. While this should be a relatively low-level effort, it is possible that technological advances may make them more practical in the future.

6. Conclusions and Recommendations

As a result of assessing ceramic materials technology for heat exchangers, several problem areas are evident and recommendations for future research can be made. This section summarizes these conclusions and recommendations in the form of a problem statement and the associated recommended research statement.

Problem--Although the currently emerging ceramic recuperators offer significant improvement over their predecessors, their durability is unpredictable.

Recommendation--In order for the designer to know the long-term durability of a proposed design, some information on long-term exposure under stress is needed. This information may be partially obtained by iterative fabrication and testing of the near-final design, but this information may only be applicable to that specific design, whereas long-term exposure and stress information for generic shapes will be the most generally applicable information to many designs.

Problem--Although candidate ceramic materials such as silicon carbide, silicon nitride, alumina, and zirconium oxide can be identified for use in the selected waste heat streams, these ceramics all exhibit characteristic ceramic brittle behavior leading to sudden complete failures. Also, variability in material properties can occur due to many different manufacturing techniques and possibly due to impurities and variation in grain sizes in powder processing.

Recommendation--Decrease the catastrophic effects of brittle behavior by toughening alkali-resistant materials such as alumina with dispersed particles or fibers.

Recommendation--Determine the cause/effect relationship of powder processing techniques to material failure by investigating techniques which decrease impurities, observing the effects on material properties.

Recommendation--Develop useful standards and specifications for manufacturing and/or procuring ceramics.

Problem--For the candidate ceramic materials, the effects of temperature, time, and corrosion on material properties is unknown.

Recommendation--Obtain time-dependent stress rupture and thermal cycle data for candidate ceramic materials.

Problem--Most of the material property data available is for bar type specimens where sample preparation has unknown effects on material behavior and extrapolation of results yields large errors.

Recommendation--Develop test methods which test a ceramic material which is in its near final form such as C-ring for tubes or determine if bar data yields a consistent error which can be factored into the extrapolation.

Problem--There is no fully developed stress-corrosion test method for ceramic materials in projected heat exchanger waste streams.

Recommendation--Test methods need to be developed which quantify these effects and provide designers with viable data relating materials behavior with specific targeted waste streams.

Problem--Current production scale product reliability techniques are only able to detect gross defects so that ceramic components which should be rejected are not rejected.

Recommendation--Develop production scale NDE techniques which have shown promise at the basic research level for detecting smaller flaws such as high frequency ultrasonics and microfocus x-ray by establishing minimum detection limits, and adapting the techniques to as-fabricated HX shapes and surface conditions.

Problem--Although there are several techniques for predicting ceramic material lifetime, these techniques have problems with correctness of failure mode assumptions and certainty of constant flaw population with time and exposure.

Recommendation--Use time-dependent stress rupture and thermal cycle data in environment to predict lifetime where a failure mode is not assumed.

Recommendation--Use NDE technology to relate stress rupture information to failure mode and to results of shorter term tests such as fast fracture.

7. References

1. P. M. Wikoff, D. J. Wiggins, R. L. Tallman, C. E. Forkel, <u>High Temperature Waste Heat Stream Selection and Characterization</u>, EGG-SE-6349, August 1983.

2. U.S. Department of Energy, <u>Ceramic Heat Recuperators for Industrial Heat Recovery</u>, Final Report, DOE/EC/02162, August 1980.

3. U.S. Department of Energy, <u>Technical Acceleration Program for High Temperature Recuperators</u>, Final Report, Contract DE-FC01-80CS40330, March 1982.

4. A. O. Hoffman, W. W. Lownie, F. C. Holden, <u>An Energy Audit of Three Energy-Conserving Devices in a Steel Industry Demonstration Program. Task III: GTE High Temperature Recuperator</u>, Final Report (Battelle Columbus Laboratories, Columbus, Ohio), January 1983.

5. G. W. Weber and V. J. Tennery, <u>Analysis of a Cordierite (MAS) Recuperator for a Molybdenum Heat-Treating Furnace</u>, ORNL/TM-6794, May 1979.

6. <u>Ceramic Heat Recuperators for Industrial Heat Recovery</u>, Interim Technical Report, DOE/CS/40174-01, May 1980.

7. J. J. Cleveland, J. M. Gonzales, K. H. Kohnken, <u>Ceramic Heat Recuperators for Industrial Heat Recovery</u>, DOE/CS-40174-T2 (GTE Products Corp.), 1980.

8. R. Dorazio et al., <u>Technology Acceleration Program for the GTE Ceramic High-Temperature Recuperator</u>, DOE Contract No. DE-FC01-80CS40330 (GTE Product Corp.), 1983.

9. J. W. Bjerklie and R. A. Penty, <u>High Temperature Recuperator Tests</u>, DOE Contract No. DE-AC0Z-79CS-40257 (Hague International), 1980.

10. H. W. Lownie and F. C. Holden, "An Energy Audit of Three Energy-Conserving Devices in a Steel-Industry Demonstration Program, Task 1," Hague Forge Furnaces, PNL-4636, February 1983.

11. M. G. Coombs, D. M. Kotchick, H. J. Strumpf, <u>High Temperature Ceramic Recuperator and Combustion Air Burner Programs</u>, GRI-82/0015 (Garrett-AiResearch), 1982

12. R. E. Womack and D. K. Stafford, <u>High Temperature Burner Duct Recuperator Program</u>, unpublished draft annual report (Babcock & Wilcox), 1982.

13. F. Rudloff, <u>High Temperature Counter Flow Recuperator</u>, Final Report, DOE/CS/12077 (Uintex Corp.), 1981.

14. M. E. Ward et al., Development of a Ceramic Tube Heat Exchanger with Relaxing Joint, FE-2556-30, Final Report (Solar Turbines International), 1980.

15. M. E. Ward and A. H. Campbell, Ceramic Heat Exchanger Technology Development, Final Report, DE-4898-F (Solar Turbines International), 1982.

16. A. Prasad and J. K. Jasti, Advanced Regenerative Heat Recovery System, Annual Report, GRI-80/0115 (Midland-Ross), 1982.

17. G. C. Wei and V. J. Tennery, "Evaluation of Tubular Ceramic Heat Exchanger Materials in Residual Oil Combustion Environment," ORNL/TM-7578, March 1981.

18. D. C. Larsen and J. W. Adams, "Property Screening and Evaluation of Ceramic Turbine Materials," Semiannual Interim Technical Report No. 12, Contract F33615-79-C-5100, IIT Research Institute, May 1982.

19. M. Coombs, D. Kotchick, and M. Weidhaas, "High Temperature Ceramic Heat Exchanger Development," AiResearch Manufacturing Company, 82-19435, Dec. 3, 1982.

20. M. E. Ward et al, "Development of a Ceramic Tube Heat Exchanger with Relaxing Joint," FE-2556-30, June 1980.

21. G. D. Quinn, "Review of Static Fatigue in Silicon Nitride and Silicon Carbide," 1981 New England Section Topical Meeting on Nonoxide Ceramics, New England Section of the American Ceramic Society, October 5-6, 1981.

22. Private Communication, J. Synder, Babcock & Wilcox, June 1983.

23. High Temperature Burner-Duct Recuperator System Progress Report for February 1983, AiResearch Manufacturing Company, 81-18493 (17).

24. G. W. Weber and V. J. Tennery, "Materials Analyses of Ceramics for Glass Furnace Recuperators," ORNL/TM-6970, November 1979.

25. M. K. Ferber and V. J. Tennery, "Evaluation of Tubular Ceramic Heat Exchanger Materials in Acidic Coal Ash from Coal-Oil-Mixture Combustion," ORNL/TM-7958, December 1981.

26. M. K. Ferber and V. J. Tennery, "Evaluation of Tubular Ceramic Heat Exchanger Materials in Basic Coal Ash from Coal-Oil-Mixture Combustion," ORNL/TM-8385, October 1982.

27. G. C. Wei, C. S. Morgan, C. R. Kennedy, D. R. Johnson, "Synthesis, Characterization, and Fabrication of Silicon Carbide Structural Ceramics," Proc. of the 7th Annual Conference on Materials for Coal Conversion and Utilization, Gaithersburg, Md., November 1982, pp. 187-219.

28. George Chia-tsing Wei, "Beta SiC Powders Produced by Carbothermic Reduction of Silica in a High-Temperature Rotary Furnace," Communications of the American Ceramic Society, July 1983, C-111 to -113.

29. S. Prochazka, "Sintered Dense Silicon Carbide," U.S. Patent 4004934, 1977.

30. S. Yajima, J. Hayashi, M. Omori, "Continuous Silicon Carbide Fiber of High Tensile Strength," Chem. Lett., 931-34 (1975).

31. C. L. Schilling, Jr., J. P. Wesson, T. C. Williams, "Organosilane Polymers, IV, Polycarbosilane Precursors for Silicon Carbides," Report 80-1, Union Carbide Technical Center, Tarrytown, NY, 1981.

32. W. R. Cannon, S. C. Danforth, J. H. Flint, J. S. Haggerty, R. A. Marra, "Sinterable Ceramic Powders from Laser-Driven Reactions: I, Process Description and Modeling," J. Am. Cer. Soc., Vol. 65, No. 7, pp. 324-330, July 1982.

33. W. R. Cannon, S. C. Danforth, J. S. Haggerty, R. A. Marra, "Sinterable Ceramic Powders from Laser-Driven Reactions: II, Powder Characteristics and Process Variables," J. Am. Cer. Soc., Vol. 65, No. 7, pp. 330-335, July 1982.

34. K. M. Prewo and J. J. Brennan, "High-Strength Silicon Carbide Fibre-Reinforced Glass-Matrix Composites," J. Mater. Sci. 15, 463-68 (1980).

35. P. F. Becher, G. C. Wei, M. K. Ferber, "Toughening Behavior in SiC Whisker Reinforced Alumina," to be published.

36. C. S. Morgan, "Thermal Shock-Resistant Cermet Insulators," NUREG/CR-2363, ORNL/TM-8038, November 1981.

37. A. J. Moorhead, P. F. Becher, R. J. Lauf, C. S. Morgan, "Fabrication, Testing and Brazing of Dispersed-Metal Toughened Alumina," Proc. 20th Automative Technology Development Contractors' Coordination Meeting P-120, Dearborn, Michigan, Oct. 25-28, 1982.

38. W. J. Lackey, ORNL, private communication to J. I. Federer, ORNL, September 1983.

39. W. D. Kingery, _Property Measurements at High Temperatures_, John Wiley and Sons, New York, 1959, p. 162.

40. R. J. H. Bollard and R. Taggart, "Mechanical Characterization," _Design of Brittle Materials_, University of Washington, J. I. Mueller, K. S. Kobayashi, W. D. Scott (eds.), Seattle, Washington, 1979, p. 4.7.

41. F. I. Baratha, _Requirements for Flexure Testing of Brittle Materials_, AMMRAC TR-82-20, April 1982.

42. M. Coombs, D. Kotchick, H. Warren, <u>High Temperature Ceramic Heat Exchanger</u>, EPRI-FP-1127, July 1979.

43. M. E. Ward, N. G. Solomon, M. E. Gulden, C. E. Smeltzer, <u>Development of a Ceramic Tube Heat Exchanger with Relaxing Joint</u>, FE-2556-30, June 1980.

44. ASTM-C 496-71, <u>Splitting Tensile Strength of Cylindrical Concrete Specimens</u>.

45. MCC-II, Splitting Tensile Standard Test Methods, Materials Char. Center Test Methods, ONL 3990, July 1981.

46. M. R. Ferber and V. J. Tennery, <u>Evaluation of Tubular Ceramic Heat Exchanger Materials in Acidic Coal Ash from Coal-0.1-Mixture Combustion</u>, ORNL-TM-7958, December 1981.

47. R. Kossowsky, "Creep, Fatigue of Si_3N_4 as Related to Microstructures," <u>Ceramic for High Performance Applications</u>, Burke, Gorum, and Katz (eds.), <u>Proceedings Sec. Army Mat. Tech. Conf., November 13-16, Watertown, Massachusetts, 1974</u>, pp. 347-372.

48. A. G. Evans, "Fracture Mechanics Determinations," <u>Fracture Mechanics of Ceramics I</u>, Brandt, Hasselman, and Lange (eds.), Plenum Press, N.Y., 1974, pp. 17-48.

49. S. M. Wiederhorn, "Subcritical Crack Growth in Ceramics," <u>Fracture Mechanics in Ceramics II</u>, Brandt, Hasselman, and Longe (eds.), Plenum Press, N.Y., pp. 613-646.

50. J. N. Hujenik and J. E. Ritter, Jr., "Susceptibility of Alumina Substrates to Stress Corrosion Cracking During Wet Processing," <u>Bull. Am. Ceram. Soc.</u>, <u>59</u>, 1981, p. 1205.

51. S. M. Wiederhorn, A. G. Evans, E. R. Fuller, and H. Johnson, "Application of Fracture Mechanics to Space-Shuttle Windows," <u>J. Am. Ceram. Soc.</u>, <u>57</u>, 1974, pp. 319-23.

52. J. E. Ritter, Jr., J. M. Sullivan, Jr., K. Jakus, "Application of Fracture Mechanics Theory to Fatigue Failure of Optical Glass Fibers," <u>J. Appl. Phys.</u>, <u>49</u>, 1978, pp. 4779-82.

53. S. W. Freiman, B. J. Hockey, S. M. Wiederhorn, "Allowable Residual Stress for Ceramic Chip Capacitors," Presented at the '83 Annual Meeting of the American Ceramic Society, Washington, D. C., May 3-5, 1981.

54. A. G. Evans and E. R. Fuller, "Proof-Testing--The Effects of Slow Crack Growth," <u>Materials Science and Engineering</u>, <u>19</u>, pp. 69-77.

55. W. Weibull, "A Statistical Theory of the Strength of Materials," Ing. Vetensk. Akad. Handl., 151, 44, 1939.

56. W. Weibull, "A Statistical Distribution Function of Wide Applicability," Journal of Applied Mechanics, 18, September 1951, pp. 293-297.

57. A. G. Evans and S. M. Wiederhorn, "Proof Testing of Ceramic Materials--An Analytical Basis for Failure Prediction," Int'l. Jour. of Fracture, 10, September 3, 1974, pp. 379-92.

58. A. G. Evans, "17. High-Temperature Slow Crack Growth in Ceramic Materials," Ceramics for High-Performance Applications, J. J. Burke, A. E. Gorum, R. N. Katz (eds.), pp. 373-96.

59. S. M. Wiederhorn, "29. Reliability, Life Prediction, and Proof Testing of Ceramics," Ceramics for High-Performance Applications, J. J. Burke, A. E. Gorum, R. N. Katz (eds.), pp. 633-63.

60. S. B. Batdorf and J. G. Crose, "A Statistical Theory for Fracture of Brittle Structures Subjected to Polyaxial Stress States," J. Appl. Mech., 41, 1974, pp. 459-65.

61. M. N. Giovan and G. Sines, "Biaxial and Uniaxial Data for Statistical Comparisons of a Ceramic's Strength," J. Am. Cer. Soc., 62, 9-10, pp. 510-15.

62. F. F. Lange, "Origin and Use of Fracture Mechanics," Fracture Mechanics of Ceramics, 1 (SAA), p. 3.

63. E. M. Lenoe, "5. Probability-Based Design and Analysis--The Reliability Problem," Ceramics for High-Performance Applications, J. J. Burke, A. E. Gorum (eds.), R. N. Katz, pp. 123-45.

64. E. R. Fuller, Jr., S. M. Wiederhorn, J. E. Ritter, Jr., P. B. Oates, "Proof Testing of Ceramics, Part I Experiment," Jour. Materials Science, 15, 1980, pp. 2275-81.

65. E. R. Fuller, Jr., et al., "Proof Testing of Ceramics, Part 2 Theory," J. Materials Science, 15, 1980, pp. 2202-95.

66. J. E. Ritter, Jr. and S. A. Wulf, "Evaluation of Proof Testing to Assure Against Delayed Failure," Ceramic Bull., 57, 2, 1978, pp. 186-89.

67. A. G. Evans, S. M. Wiederhorn, M. Lineer, E. R. Fuller, Jr., "Proof Testing of Porcelain Insulators and Application of Acoustic Emission," Am. Ceramic Society Ceramic Bulletin, 54, 6, 1975, pp. 576-81.

68. J. E. Ritter, Jr., K. Jakus, L. A. Strazdis, W. D. Rogers, "Effect of High Temperature Exposure in Air on Strength of Hot-Pressed Silicon Nitride," J of Am. Cer. Soc., 66, 3, March 1983, C-53-55.

69. S. M. Wiederhorn, N. J. Tighe, "Proof Testing of Hot-Pressed Silicon Nitride," J. Material Science, 13, 1978, pp. 1781-93.

70. L. P. Wynn, D. J. Tree, T. M. Yonushonis, R. A. Solomon, "Proof Testing of Ceramic Components," pp. 493-516.

71. S. Sakaguchi and M. Nakahara, "Strength of Proof Test Optical Fibers," Communications An. Cor. S, JACTAW, __, 3, 1963, C-46-47.

72. R. J. H. Bollard, "Unit 3, Theory of Elasticity," Design with Brittle Materials, J. I. Muller, A. S. Kobayashi, W. D. Scot (eds.), Univ. of Washington, Seattle, 1979.

73. G. G. Tratina, "Strength and Proof Testing of SiC," J of Materials Science, 17, 1982, pp. 1487-92.

74. G. Quinn, "Characterization of Turbine Ceramics after Long-Term Environmental Exposure," AMMRAC TR 8-15, April 1980.

75. G. Quinn and R. Latz, "Time-Dependent High Temperature Strength of Sintered α-SiC," J. Am. Ceram. Soc., 63, 1-2, 1980, pp. 117-9.

76. R. Rice, S. Freiman, J. Mechjolshy, R. Ruh, Y. Harada, in Ceramics for High Performance Applications, II, J. Burke, E. Lenoe, and R. Katz (eds.), Brook Hill, MA, 1978, pp. 669-88.

77. O. R. Gericke, "Review of Nondestructive Inspection Techniques," Fracture Mechanics of Ceramics, Vol. I, R. C. Brandt, D. P. H. Hasselman, F. Lange (eds.), New York-London, Plenum Press, 1974, pp. 117-160.

78. John L. Bjorkstam, "Nondestructive Evaluation, Unit 9," Design with Brittle Materials, J. L. Mueller, A. S. Kobayashi, W. D. Scott (eds.), Seattle, University of Washington, 1979.

79. P. K. Kandelval, P. W. Heitman, A. J. Silversmith, T. D. Wakefield, "Surface Flaw Detection in Structural Ceramics by Scanning Photoacoustic Spectroscopy," Applied Physics Letters, 37, 9, November 1980.

80. C. D. McGillem, C. L. Chen, E. S. Furgason, R. L. Gunshor, A. C. Kak and V. L. Newhouse, NASA Report No. CR-159865 (unpublished 1980).

81. G. A. McDaniel, "Recent Developments in High Output Microfocus X-Ray Systems," Proceedings of Automated Nondestructive Testing Seminar, Idaho Falls, Idaho, June 28-30, 1983, pp. 160-162.

82. R. K. Mueller, R. L. Rylander, "'Seeing' Acoustically," IEEE Spectrum, February 2, 1982, pp. 28-33.

83. D. S. Kupperman and D. Yuhas, <u>Development of Nondestructive Evaluation Techniques for High Temperature Ceramic Heat Exchanger Components</u>, ANL/FE-82-2, September 1981.

84. D. S. Kupperman, M. J. Caines, A. Winiecki, "Assessment of Ultrasonic NDE Methods for Ceramic Heat Exchanger Tubes," <u>Materials Evaluation</u>, $\underline{40}$, 7, June 1982, p. 774.

85. T. Derkacs, I. M. Matay, W. D. Brentnall, <u>Nondestructive Evaluation of Ceramics</u>, TRW Technical Report ER-7798-F, May 1976.

86. J. J. Schuldies, "The Acoustic Emission Response of Mechanically Stressed Ceramics," <u>Materials Evaluation</u>, $\underline{31}$, 10, 1973, p. 209.

87. A. G. Evans and M. Linzer, "Failure Prediction in Structural Ceramics Using Acoustic Emission," <u>Journal of the American Ceramic Society</u>, $\underline{56}$, 11, November 1973.

88. J. Gonazalez, private communication, GTE-Sylvania, Towanda, Pennsylvania, April 19, 1983.

89. D. Kotchick, private communication, Garrett AiResearch, Torrance, CA, May 4, 1983.

Appendix A

Ceramic Materials Properties Data

This appendix contains tables of properties data, obtained from manufacturers, on various ceramic materials of interest for possible use in designing and fabricating waste heat recuperators. Conversion factors for SI units are given in Appendix B.

Appendix A: Ceramic Materials Properties Data   353

TABLE A-1. PROPERTIES OF SILICON CARBIDE

IDENTIFICATION

Manufacturer -- The Carborundum Company
Designation -- Sintered Alpha (Sintered-α)
Nominal Composition -- >98 wt % SiC

PROPERTY	UNITS	TEMPERATURE °C								
		25	200	400	600	800	1000	1200	1400	1600
A. PHYSICAL										
1 Hardness (Knoop)	Kg/mm^2	2800								
2 Bulk Density	Mg/m^3	3.14-3.18								
3 Porosity	vol. %									
4 Permeability	Darcy									
5 Specific Heat	J/kg·K	670	921	1055	1122					
6 Thermal Conductivity	W/m·K	87.0	66.9	61.5	49.4					
7 Thermal Diffusivity	cm^2/s	0.413	0.230	0.185	0.140					
8 Thermal Expansion Coefficient	10^{-6}/°C	4.02(25-700°C); 5.32(700-2000°C)								
9 Maximum Use Temperature	°C									
10 Thermal Shock Resistance										
B. MECHANICAL										
1 Flexure Strength (in air)^{a,b}	MPa	459					442	450	432	404(1500°C)
2 Weibull Modulus^c		12.3								
3 Tensile Strength	MPa									
4 Fracture Toughness^d	MPa·m$^{1/2}$	4.6					6.4			
5 Compressive Strength	MPa									
6 Young's Modulus	GPa	406					378			1398(1500°C)
7 Bulk Modulus	GPa									
8 Shear Modulus	GPa	178					169			
9 Poisson Ratio		0.142					0.118			
C. CHEMICAL										
1 Oxidation Rate Constant	g^2/cm^4·s								Not detectable	
D. ELECTRICAL										
1 Dielectric Constant										
2 Dielectric Strength	ac-kV/mm									
3 Loss Tangent										
4 Volume Resistivity	ohm·cm									

TABLE A-1. (continued)

PROPERTY	UNITS	TEMPERATURE °C								
		25	200	400	600	800	1000	1200	1400	1600
E. OPTICAL										
1 Reflectivity										
2 Spectral Emissivity										
3 Total Emissivity										

a. 4-point loading.
b. Values in argon: 446 MPa at 25; 494 MPa at 1650°C.
c. 2-parameter.
d. Double torsion and single edge notched beam.

Appendix A: Ceramic Materials Properties Data   355

TABLE A-2. PROPERTIES OF SILICONIZED SILICON CARBIDE

IDENTIFICATION

Manufacturer -- The Carborundum Company
Designation -- Hexalloy KT Fine Grained Reaction Sintered
Nominal Composition -- 84 wt% SiC, 16 wt % Si

PROPERTY	UNITS	TEMPERATURE °C								
		25	200	400	600	800	1000	1200	1400	1600
A. PHYSICAL										
1 Hardness (Knoop)	kg/mm^2	1860								
2 Bulk Density	Mg/m^3	2.97-2.99								
3 Porosity	vol. %									
4 Permeability	darcy									
5 Specific Heat	J/kg·K	711			1090	1170	1260	1340	1420	
6 Thermal Conductivity	W/m·K	104.6			38.9	33.1	31.8	31.8	29.7	
7 Thermal Diffusivity	cm^2/s	0.49			0.120	0.095	0.085	0.080	0.070	
8 Thermal Expansion Coefficient	10^{-6}/°C	3.44(25-500°C); 5.02(500-1400°C)								
9 Maximum Use Temperature	°C									
10 Thermal Shock Resistance										
B. MECHANICAL										
1 Flexure Strength (in air)[a]	MPa	383			408	441	461	491	378	
2 Weibull Modulus[b]		10					16		19	
3 Tensile Strength	MPa									
4 Fracture Toughness[c]	MPa·m$^{1/2}$	4.97					5.04	5.74	7.66	
5 Compressive Strength	MPa	332								
6 Young's Modulus	GPa									
7 Bulk Modulus	GPa									
8 Shear Modulus	GPa	147								
9 Poisson Ratio		0.127								
C. CHEMICAL										
1 Oxidation Rate Constant	g^2/cm^4·s						1.42×10^{-14} (1150°C)	4.06×10^{-14} (1260°C)	7.70×10^{-14} (1370°C)	
D. ELECTRICAL										
1 Dielectric Constant										
2 Dielectric Strength	ac-kV/mm									
3 Loss Tangent										
4 Volume Resistivity	ohm·cm									

TABLE A-2. (continued)

PROPERTY	UNITS	TEMPERATURE °C								
		25	200	400	600	800	1000	1200	1400	1600
E. OPTICAL										
1 Reflectivity										
2 Spectral Emissivity										
3 Total Emissivity										

a. 4-point loading.
b. 2-parameter.
c. Single edge notched beam.

Appendix A: Ceramic Materials Properties Data 357

TABLE A-3. PROPERTIES OF SILICONIZED SILICON CARBIDE

IDENTIFICATION

Manufacturer -- Coors Porcelain Company
Designation -- Reaction Sintered SC-2
Nominal Composition -- 88 wt % SiC, 12 wt % Si

PROPERTY	UNITS	TEMPERATURE °C								
		25	200	400	600	800	1000	1200	1400	1600
A. PHYSICAL										
1 Hardness (Knoop)	Kg/mm^2	94R$_3$0N[a]								
2 Bulk Density	Mg/m^3	3.10								
3 Porosity	vol. %									
4 Permeability	Darcy									
5 Specific Heat	J/kg·K									
6 Thermal Conductivity	W/m·K									
7 Thermal Diffusivity	cm^2/s									
8 Thermal Expansion Coefficient	10^{-6}/°C	4.3 (25-1000°C)								
9 Maximum Use Temperature	°C									
10 Thermal Shock Resistance										
B. MECHANICAL										
1 Flexure Strength (in air)[b]	MPa	517								
2 Weibull Modulus										
3 Tensile Strength	MPa	307								
4 Fracture Toughness	MPa·m$^{1/2}$									
5 Compressive Strength	MPa	2480								
6 Young's Modulus	GPa	395								
7 Bulk Modulus	GPa									
8 Shear Modulus	GPa									
9 Poisson Ratio		0.19								
C. CHEMICAL										
1 Oxidation Rate Constant	g^2/cm^4·s									
D. ELECTRICAL										
1 Dielectric Constant										
2 Dielectric Strength	ac-kV/mm									
3 Loss Tangent										
4 Volume Resistivity	ohm·cm									

TABLE A-3. (continued)

PROPERTY	UNITS	TEMPERATURE °C								
		25	200	400	600	800	1000	1200	1400	1600
E. OPTICAL										
1 Reflectivity										
2 Spectral Emissivity										
3 Total Emissivity										

a. Knoop hardness not available.
b. 3-point loading.

Appendix A: Ceramic Materials Properties Data 359

TABLE A-4. PROPERTIES OF SILICONIZED SILICON CARBIDE

IDENTIFICATION

Manufacturer -- Norton Company
Designation -- Noralide NC430
Nominal Composition -- 90 wt % SiC, 10 wt % Si

PROPERTY	UNITS	\multicolumn{9}{c}{TEMPERATURE °C}
		25	200	400	600	800	1000	1200	1400	1600

A. PHYSICAL

1 Hardness of Major Crystal (Knoop)	Kg/mm^2									
2 Bulk Density	Mg/m^3	3.1								
3 Porosity	vol. %									
4 Permeability	Darcy									
5 Specific Heat	J/kg·K									
6 Thermal Conductivity	W/m·K	140.0								
7 Thermal Diffusivity	cm^2/s									
8 Thermal Expansion Coefficient	10^{-6}/°C	4.8								
9 Maximum Use Temperature	°C									
10 Thermal Shock Resistance										

B. MECHANICAL

1 Flexure Strength (in air)	MPa	310								
2 Weibull Modulus										
3 Tensile Strength	MPa									
4 Fracture Toughness	MPa·m$^{1/2}$	3.5								
5 Compressive Strength	MPa									
6 Young's Modulus	GPa	400								
7 Bulk Modulus	GPa									
8 Shear Modulus	GPa									
9 Poisson Ratio										

C. CHEMICAL

1 Oxidation Rate Constant	g^2/cm^4·s									

D. ELECTRICAL

1 Dielectric Constant										
2 Dielectric Strength	ac-kV/mm									
3 Loss Tangent										
4 Volume Resistivity	ohm·cm									

TABLE A-4. (continued)

PROPERTY	UNITS	TEMPERATURE °C								
		25	200	400	600	800	1000	1200	1400	1600

E. OPTICAL

1 Reflectivity
2 Spectral Emissivity
3 Total Emissivity

Appendix A: Ceramic Materials Properties Data 361

TABLE A-5. PROPERTIES OF SILICONIZED SILICON CARBIDE

IDENTIFICATION

Manufacturer -- Norton Company
Designation -- Crystar-HD
Nominal Composition -- 89 wt % SiC (99% pure), 11 wt % Si

| | PROPERTY | UNITS | \multicolumn{9}{c}{TEMPERATURE °C} |
			25	200	400	600	800	1000	1200	1400	1600
A.	**PHYSICAL**										
1	Hardness	Kg/mm²	2700[a]								
2	Bulk Density	Mg/m³	3.1								
3	Porosity	vol. %	0								
4	Permeability	Darcy									
5	Specific Heat	cal./g·°C									
6	Thermal Conductivity	W/m·K					50 (650)				
7	Thermal Diffusivity	cm²/s									
8	Thermal Expansion Coefficient	10⁻⁶/°C	5.0 (30-1500°C)								
9	Maximum Use Temperature	°C								1370	
10	Thermal Shock Resistance										
B.	**MECHANICAL**										
1	Flexure Strength (in air)	MPa	300								
2	Weibull Modulus										
3	Tensile Strength	MPa									
4	Fracture Toughness	MPa·m^(1/2)									
5	Compressive Strength	MPa									
6	Young's Modulus	GPa									
7	Bulk Modulus	GPa									
8	Shear Modulus	GPa									
9	Poisson Ratio										
C.	**CHEMICAL**										
1	Oxidation Rate Constant	g²/cm⁴·s									
D.	**ELECTRICAL**										
1	Dielectric Constant										
2	Dielectric Strength	ac-kV/mm									
3	Loss Tangent										
4	Volume Resistivity	ohm·cm									

TABLE A-5. (continued)

PROPERTY	UNITS	TEMPERATURE °C									
		25	200	400	600	800	1000	1200	1400	1600
E. OPTICAL										
1 Reflectivity										
2 Spectral Emissivity										
3 Total Emissivity		0.95								

a. Principal crystal.

Appendix A: Ceramic Materials Properties Data 363

TABLE A-6. PROPERTIES OF SILICONIZED SILICON CARBIDE

IDENTIFICATION

Manufacturer -- Norton Company
Designation -- Crystar-K
Nominal Composition -- 85 wt % SiC (99%+ pure), 15 wt % Si

PROPERTY	UNITS	\multicolumn{9}{c}{TEMPERATURE °C}								
		25	200	400	600	800	1000	1200	1400	1600

A. PHYSICAL

1 Hardness of Major Crystal (Knoop)	Kg/mm^2									
2 Bulk Density	Mg/m^3	3.0								
3 Porosity	vol. %									
4 Permeability	Darcy									
5 Specific Heat	J/kg·K									
6 Thermal Conductivity	W/m·K				1130 (727)					
7 Thermal Diffusivity	cm^2/s				38 (727)					
8 Thermal Expansion Coefficient	10^{-6}/°C	4.8 (25 to 1000°C)								
9 Maximum Use Temperature	°C								1400	
10 Thermal Shock Resistance										

B. MECHANICAL

1 Flexure Strength (in air)	MPa							172		
2 Weibull Modulus										
3 Tensile Strength	MPa									
4 Fracture Toughness	MPa·m$^{1/2}$									
5 Compressive Strength	MPa	280								
6 Young's Modulus	GPa									
7 Bulk Modulus	GPa									
8 Shear Modulus	GPa									
9 Poisson Ratio										

C. CHEMICAL

1 Oxidation Rate Constant	g^2/cm^4·s									

D. ELECTRICAL

1 Dielectric Constant										
2 Dielectric Strength	ac-kV/mm									
3 Loss Tangent		0.1								
4 Volume Resistivity	ohm·cm									

TABLE A-6. (continued)

PROPERTY	UNITS	TEMPERATURE °C									
		25	200	400	600	800	1000	1200	1400	1600
E. OPTICAL										
1 Reflectivity										
2 Spectral Emissivity										
3 Total Emissivity										

Appendix A: Ceramic Materials Properties Data 365

TABLE A-7. PROPERTIES OF RECRYSTALLIZED SILICON CARBIDE

IDENTIFICATION

Manufacturer -- Norton Company
Designation -- Crystar
Nominal Composition -- 99+ wt % SiC

PROPERTY	UNITS	TEMPERATURE °C									
		25	200	400	600	800	1000	1200	1400	1600
A. PHYSICAL										
1 Hardness	kg/mm^2									
2 Bulk Density	Mg/m^3	2.6								
3 Porosity	vol. %									
4 Permeability	Darcy									
5 Specific Heat	$J/kg \cdot K$									
6 Thermal Conductivity	$W/m \cdot K$					1170(727)				
7 Thermal Diffusivity	cm^2/s					24(727)				
8 Thermal Expansion Coefficient	$10^{-6}/°C$	4.8 (25 to 1000°C)								
9 Maximum Use Temperature	°C									1600
10 Thermal Shock Resistance										
B. MECHANICAL										
1 Flexure Strength (in air)	MPa							117		
2 Weibull Modulus										
3 Tensile Strength	MPa									
4 Fracture Toughness	$MPa \cdot m^{1/2}$									
5 Compressive Strength	MPa									
6 Young's Modulus	GPa	210								
7 Bulk Modulus	GPa									
8 Shear Modulus	GPa									
9 Poisson Ratio	GPa									
C. CHEMICAL										
1 Oxidation Rate Constant	$g^2/cm^4 \cdot s$									
D. ELECTRICAL										
1 Dielectric Constant										
2 Dielectric Strength	ac-kV/mm									
3 Loss Tangent					0.3					
4 Volume Resistivity	ohm·cm	100								

TABLE A-7. (continued)

PROPERTY	UNITS	TEMPERATURE °C								
		25	200	400	600	800	1000	1200	1400	1600
E. OPTICAL										
1 Reflectivity										
2 Spectral Emissivity										
3 Total Emissivity										

Appendix A: Ceramic Materials Properties Data 367

TABLE A-8. PROPERTIES OF NITRIDE-BONDED SILICON CARBIDE

IDENTIFICATION

Manufacturer -- Coors Porcelain Company
Designation -- Cerasurf TN-15
Nominal Composition -- 66 wt % SiC, 32 wt % Si_3N_4, 2 wt % oxides

PROPERTY	UNITS	TEMPERATURE °C								
		25	200	400	600	800	1000	1200	1400	1600
A. PHYSICAL										
1 Hardness	Kg/mm²	2700[a]								
2 Bulk Density	Mg/m³	2.65								
3 Porosity	vol. %	14								
4 Permeability	Darcy									
5 Specific Heat	J/kg·K									
6 Thermal Conductivity	W/m·K						15.9 (982°C)			
7 Thermal Diffusivity	cm²/s									
8 Thermal Expansion Coefficient	10^{-6}/°C									
9 Maximum Use Temperature	°C										1590
10 Thermal Shock Resistance										
B. MECHANICAL										
1 Flexure Strength (in air)	MPa	46.2								
2 Weibull Modulus										
3 Tensile Strength	MPa									
4 Fracture Toughness	MPa·m$^{1/2}$									
5 Compressive Strength	MPa									
6 Young's Modulus	GPa									
7 Bulk Modulus	GPa									
8 Shear Modulus	GPa									
9 Poisson Ratio										
C. CHEMICAL										
1 Oxidation Rate Constant	q²/cm⁴·s									
D. ELECTRICAL										
1 Dielectric Constant										
2 Dielectric Strength	ac-kV/mm									
3 Loss Tangent										
4 Volume Resistivity	ohm·cm									

TABLE A-8. (continued)

PROPERTY	UNITS	\multicolumn{8}{c}{TEMPERATURE °C}
		25	200	400	600	800	1000	1200	1400	1600

E. OPTICAL
1 Reflectivity
2 Spectral Emissivity
3 Total Emissivity

a. Principal crystal.

Appendix A: Ceramic Materials Properties Data   369

TABLE A-9. PROPERTIES OF NITRIDE BONDED SILICON CARBIDE

IDENTIFICATION

Manufacturer -- Norton Company
Designation -- Cryston CN178
Nominal Composition -- 75 wt % SiC, 23 wt % Si_3N_4, 2 wt % Oxides

PROPERTY	UNITS	TEMPERATURE °C								
		25	200	400	600	800	1000	1200	1400	1600
A. PHYSICAL										
1 Hardness	Kg/mm^2	2700^a								
2 Bulk Density	Mg/m^3	2.6								
3 Porosity	vol. %	18								
4 Permeability	Darcy									
5 Specific Heat	J/kg·K									
6 Thermal Conductivity	W/m·K				17 (650)					
7 Thermal Diffusivity	cm^2/s									
8 Thermal Expansion Coefficient	10^{-6}/°C	5.0 (30-1500°C)								
9 Maximum Use Temperature	°C									1590
10 Thermal Shock Resistance										
B. MECHANICAL										
1 Flexure Strength (in air)	MPa	59								
2 Weibull Modulus										
3 Tensile Strength	MPa									
4 Fracture Toughness	MPa·$m^{1/2}$									
5 Compressive Strength	MPa									
6 Young's Modulus	GPa									
7 Bulk Modulus	GPa									
8 Shear Modulus	GPa									
9 Poisson Ratio										
C. CHEMICAL										
1 Oxidation Rate Constant	g^2/cm^4·s									
D. ELECTRICAL										
1 Dielectric Constant										
2 Dielectric Strength	ac-kV/mm									
3 Loss Tangent										
4 Volume Resistivity	ohm·cm									

TABLE A-9. (continued)

PROPERTY	UNITS	TEMPERATURE °C								
		25	200	400	600	800	1000	1200	1400	1600
E. OPTICAL										
1 Reflectivity										
2 Spectral Emissivity										
3 Total Emissivity		0.92								

a. Principal crystal.

Appendix A: Ceramic Materials Properties Data 371

TABLE A-10. PROPERTIES OF NITRIDE BONDED SILICON CARBIDE

IDENTIFICATION

Manufacturer -- Norton Company
Designation -- Cryston CN983
Nominal Composition -- 78 wt % SiC, 22 wt % Si_3N_4

PROPERTY	UNITS	TEMPERATURE °C								
		25	200	400	600	800	1000	1200	1400	1600
A. PHYSICAL										
1 Hardness	Kg/mm^2									
2 Bulk Density	Mg/m^3	2.50								
3 Porosity	vol. %	18								
4 Permeability	Darcy									
5 Specific Heat	J/kg·K									
6 Thermal Conductivity	W/m·K		15.6 (315)				14.7 (650)	13.3 (982)	12.7 (1149)	
7 Thermal Diffusivity	cm^2/s									
8 Thermal Expansion Coefficient	10^{-6}/°C	4.7 (30-1500°C)								
9 Maximum Use Temperature	°C									1540
10 Thermal Shock Resistance										
B. MECHANICAL										
1 Flexure Strength (in air)	MPa	100							90 (1450)	
2 Weibull Modulus										
3 Tensile Strength	MPa									
4 Fracture Toughness	MPa·m$^{1/2}$									
5 Compressive Strength	MPa									
6 Young's Modulus	GPa									
7 Bulk Modulus	GPa									
8 Shear Modulus	GPa									
9 Poisson Ratio										
C. CHEMICAL										
1 Oxidation Rate Constant	g^2/cm^4·s									
D. ELECTRICAL										
1 Dielectric Constant										
2 Dielectric Strength	ac-kV/mm									
3 Loss Tangent										
4 Volume Resistivity	ohm·cm									

TABLE A-10. (continued)

PROPERTY	UNITS	TEMPERATURE °C								
		25	200	400	600	800	1000	1200	1400	1600

E. OPTICAL

1 Reflectivity
2 Spectral Emissivity
3 Total Emissivity

Appendix A: Ceramic Materials Properties Data 373

TABLE A-11. PROPERTIES OF OXINITRIDE BONDED SILICON CARBIDE

IDENTIFICATION

Manufacturer -- Norton Company
Designation -- Crystolon CN163
Nominal Composition -- 85 wt % SiC, 13 wt % Si_2ON_2, 2 wt % Oxides

PROPERTY	UNITS	TEMPERATURE °C								
		25	200	400	600	800	1000	1200	1400	1600
A. PHYSICAL										
1 Hardness	Kg/mm^2	2700[a]								
2 Bulk Density	Mg/m^3	2.6								
3 Porosity	vol. %	15								
4 Permeability	Darcy									
5 Specific Heat	J/kg·K									
6 Thermal Conductivity	W/m·K					18 (650)				
7 Thermal Diffusivity	cm^2/s									
8 Thermal Expansion Coefficient	$10^{-6}/°C$	5.0 (30-1500°C)								
9 Maximum Use Temperature	°C									1590
10 Thermal Shock Resistance										
B. MECHANICAL										
1 Flexure Strength (in air)	MPa	24								
2 Weibull Modulus										
3 Tensile Strength	MPa									
4 Fracture Toughness	MPa·$m^{1/2}$									
5 Compressive Strength	MPa									
6 Young's Modulus	GPa									
7 Bulk Modulus	GPa									
8 Shear Modulus	GPa									
9 Poisson Ratio										
C. CHEMICAL										
1 Oxidation Rate Constant	$g^2/cm^4 \cdot s$									
D. ELECTRICAL										
1 Dielectric Constant										
2 Dielectric Strength	ac-kV/mm									
3 Loss Tangent										
4 Volume Resistivity	ohm·cm									

TABLE A-11. (continued)

PROPERTY	UNITS	\multicolumn{8}{c}{TEMPERATURE °C}
		25	200	400	600	800	1000	1200	1400	1600
E. OPTICAL										
1 Reflectivity										
2 Spectral Emissivity		0.94								
3 Total Emissivity										

a. Principal crystal.

Appendix A: Ceramic Materials Properties Data 375

TABLE A-12. PROPERTIES OF OXIDE BONDED SILICON CARBIDE

IDENTIFICATION

Manufacturer -- Norton Company
Designation -- Crystolon CN181
Nominal Composition -- 70 wt % SiC, 30 wt % Oxides

PROPERTY	UNITS	TEMPERATURE °C								
		25	200	400	600	800	1000	1200	1400	1600
A. PHYSICAL										
1 Hardness	Kg/mm^2	2700[a]								
2 Bulk Density	Mg/m^3	2.6								
3 Porosity	vol. %	18								
4 Permeability	Darcy									
5 Specific Heat	J/kg·K									
6 Thermal Conductivity	W/m·K				10 (650)					
7 Thermal Diffusivity	cm^2/s									
8 Thermal Expansion Coefficient	10^{-6}/°C	5.0 (30-1500°C)								
9 Maximum Use Temperature	°C							1260		
10 Thermal Shock Resistance										
B. MECHANICAL										
1 Flexure Strength (in air)	MPa	29								
2 Weibull Modulus										
3 Tensile Strength	MPa									
4 Fracture Toughness	MPa·m$^{1/2}$									
5 Compressive Strength	MPa									
6 Young's Modulus	GPa									
7 Bulk Modulus	GPa									
8 Shear Modulus	GPa									
9 Poisson Ratio										
C. CHEMICAL										
1 Oxidation Rate Constant	g^2/cm^4·s									
D. ELECTRICAL										
1 Dielectric Constant										
2 Dielectric Strength	ac-kV/mm									
3 Loss Tangent										
4 Volume Resistivity	ohm·cm									

TABLE A-12. (continued)

PROPERTY	UNITS	TEMPERATURE °C								
		25	200	400	600	800	1000	1200	1400	1600
E. OPTICAL										
1 Reflectivity										
2 Spectral Emissivity										
3 Total Emissivity										

a. Principal crystal.

Appendix A: Ceramic Materials Properties Data 377

TABLE A-13. PROPERTIES OF ALUMINUM OXIDE

IDENTIFICATION

Manufacturer -- Coors Porcelain Company
Designation -- AD-999
Nominal Composition -- 99.9 wt % Al_2O_3

PROPERTY	UNITS	TEMPERATURE °C								
		25	200	400	600	800	1000	1200	1400	1600
A. PHYSICAL										
1 Hardness	Kg/mm^2	1550[a]								
2 Bulk Density	Mg/m^3	3.96								
3 Porosity	vol. %	None[b]								
4 Permeability	Darcy	880								
5 Specific Heat	J/Kg·K	38.9								
6 Thermal Conductivity	W/m·K		27.6(100°C)	13.4		6.3				
7 Thermal Diffusivity	cm^2/s									
8 Thermal Expansion Coefficient	$10^{-6}/°C$	6.5(25-200°C); 7.4(25-500°C); 7.8(25-800°C); 8.0(25-1000°C); 8.3(25-1200°C)								
9 Maximum Use Temperature	°C									1900
10 Thermal Shock Resistance										
B. MECHANICAL										
1 Flexure Strength (in air)	MPa	517[c]					379[c]			
2 Weibull Modulus										
3 Tensile Strength	MPa	310					221			
4 Fracture Toughness	$MPa·m^{1/2}$									
5 Compressive Strength	MPa	3792					1930			
6 Young's Modulus	GPa	386								
7 Bulk Modulus	GPa	228								
8 Shear Modulus	GPa	158								
9 Poisson Ratio		0.22								
C. CHEMICAL										
1 Oxidation Rate Constant	$g^2/cm^4·s$									
D. ELECTRICAL										
1 Dielectric Constant		9.9[d]								
2 Dielectric Strength	ac-kV/mm	9.4[e]								
3 Loss Tangent	ohm·cm				3.3x10^{12}(500°C)		1.1x10^7			
4 Volume Resistivity	ohm·cm	>10^{15}								

TABLE A-13. (continued)

PROPERTY	UNITS	TEMPERATURE °C							
		25	200	400	600	800	1000	1200	1400	1600

E. OPTICAL

1. Reflectivity
2. Spectral Emissivity
3. Total Emissivity

a. 1000-g load.
b. No He leak through a plate 25.4 mm diam by 0.25 mm thick at 133 Pa vacuum vs. 1×10^5 Pa of He pressure for 15 s at room temperature.
c. Mean value for 100 specimens.
d. At 1 kHz; 9.8 at 1 MHz.
e. 6.35 mm thick specimen.

Appendix A: Ceramic Materials Properties Data 379

TABLE A-14. PROPERTIES OF ALUMINUM OXIDE

IDENTIFICATION

Manufacturer -- Coors Porcelain Company
Designation -- AD-995
Nominal Composition -- 99.5 wt % Al_2O_3

PROPERTY	UNITS	TEMPERATURE °C								
		25	200	400	600	800	1000	1200	1400	1600
A. PHYSICAL										
1 Hardness	Kg/mm²	1500[a]								
2 Bulk Density	Mg/m³	3.89								
3 Porosity	vol. %	None[b]								
4 Permeability	Darcy									
5 Specific Heat	J/kg·K	880								
6 Thermal Conductivity	W/m·K	35.6	25.9 (100°C)	12.1			6.3			
7 Thermal Diffusivity	cm²/s									
8 Thermal Expansion Coefficient	10^{-6}/°C	7.1(25-200°C); 7.6(25-500°C); 8.0(25-800°C); 8.3(25-1000°C)								
9 Maximum Use Temperature	°C									1750
10 Thermal Shock Resistance										
B. MECHANICAL										
1 Flexure Strength (in air)	MPa	379								
2 Weibull Modulus										
3 Tensile Strength	MPa	262								
4 Fracture Toughness	MPa·m^{1/2}									
5 Compressive Strength	MPa	2620								
6 Young's Modulus	GPa	372								
7 Bulk Modulus	GPa	228								
8 Shear Modulus	GPa	152								
9 Poisson Ratio		0.22								
C. CHEMICAL										
1 Oxidation Rate Constant	g²/cm⁴·s									
D. ELECTRICAL										
1 Dielectric Constant		9.8[c]								
2 Dielectric Strength	ac-kV/mm	8.7[d]								
3 Loss Tangent										
4 Volume Resistivity	ohm·cm	>10^{14}								

TABLE A-14. (continued)

PROPERTY	UNITS	TEMPERATURE °C								
		25	200	400	600	800	1000	1200	1400	1600

E. OPTICAL

1 Reflectivity
2 Spectral Emissivity
3 Total Emissivity

a. 1000-g load.
b. No He leak through a plate 25.4 mm diam by 0.25 mm thick at 133 Pa vacuum vs. 1×10^5 Pa of He pressure for 15 s at room temperature.
c. At 1 kHz; 9.7 at 1 MHz.
d. 6.35 mm thick specimen.

Appendix A: Ceramic Materials Properties Data   381

TABLE A-15. PROPERTIES OF MULLITE

IDENTIFICATION

Manufacturer -- Coors Porcelain Company
Designation -- Mullite
Nominal Composition -- 72 wt % Al_2O_3, 28 wt % SiO_2 (equivalent oxides)

PROPERTY	UNITS	TEMPERATURE °C								
		25	200	400	600	800	1000	1200	1400	1600
A. PHYSICAL										
1 Hardness	Kg/mm^2									
2 Bulk Density	Mg/m^3	2.82								
3 Porosity	vol. %									
4 Permeability	Darcy	gas tight								
5 Specific Heat	$J/kg \cdot K$	848(100°C)								
6 Thermal Conductivity	$W/m \cdot K$	4.1	3.9(100°C)							
7 Thermal Diffusivity	cm^2/s									
8 Thermal Expansion Coefficient	$10^{-6}/°C$	3.7(25-200°C); 4.2(25-500°C); 4.8(25-800°C); 5.0(25-1000°C)								
9 Maximum Use Temperature	°C									1700
10 Thermal Shock Resistance										
B. MECHANICAL										
1 Flexure Strength (in air)	MPa	186					151			
2 Weibull Modulus[c]										
3 Tensile Strength	MPa									
4 Fracture Toughness[d]	$MPa \cdot m^{1/2}$									
5 Compressive Strength	MPa	551								
6 Young's Modulus	GPa	155								
7 Bulk Modulus	GPa									
8 Shear Modulus	GPa									
9 Poisson Ratio	GPa									
C. CHEMICAL										
1 Oxidation Rate Constant	$g^2/cm^4 \cdot s$									
D. ELECTRICAL										
1 Dielectric Constant										
2 Dielectric Strength	ac-kV/mm	9.8[a]								
3 Loss Tangent										
4 Volume Resistivity	ohm·cm	$>10^{13}$	1×10^{12} (300)	2×10^7 (700)	2×10^7 (700)		3×10^5			

TABLE A-15. (continued)

PROPERTY	UNITS	TEMPERATURE °C								
		25	200	400	600	800	1000	1200	1400	1600
E. OPTICAL										
1 Reflectivity										
2 Spectral Emissivity										
3 Total Emissivity										

a. 6.36 mm thick.

Appendix A: Ceramic Materials Properties Data   383

TABLE A-16. PROPERTIES OF CORDIERITE

IDENTIFICATION

Manufacturer -- Coors Porcelain Company
Designation -- CD-1
Nominal Composition -- 14 wt % MgO, 35 wt % Al_2O_3, 51 wt % SiO_2 (equivalent oxides)

PROPERTY	UNITS	TEMPERATURE °C								
		25	200	400	600	800	1000	1200	1400	1600
A. PHYSICAL										
1 Hardness	Kg/mm^2	740								
2 Bulk Density	Mg/m^3	2.5								
3 Porosity	vol. %									
4 Permeability	Darcy									
5 Specific Heat	J/kg·K									
6 Thermal Conductivity	W/m·K	4.12								
7 Thermal Diffusivity	cm^2/s									
8 Thermal Expansion Coefficient	$10^{-6}/°C$	106(25-320°C)								
9 Maximum Use Temperature	°C									
10 Thermal Shock Resistance										
B. MECHANICAL										
1 Flexure Strength (in air)[a,b]	MPa									
2 Weibull Modulus[c]										
3 Tensile Strength	MPa	124								
4 Fracture Toughness[d]	MPa·$m^{1/2}$									
5 Compressive Strength	MPa									
6 Young's Modulus	GPa	138								
7 Bulk Modulus	GPa									
8 Shear Modulus	GPa									
9 Poisson Ratio		0.316								
C. CHEMICAL										
1 Oxidation Rate Constant	$g^2/cm^4·s$									
D. ELECTRICAL										
1 Dielectric Constant		4.86[a]								
2 Dielectric Strength	kV/mm	9.1[a]								
3 Loss Tangent		0.0013[a]								
4 Volume Resistivity	ohm·cm									

TABLE A-16. (continued)

PROPERTY	UNITS	TEMPERATURE °C								
		25	200	400	600	800	1000	1200	1400	1600
E. OPTICAL										
1 Reflectivity										
2 Spectral Emissivity										
3 Total Emissivity										

a. 8-10 GHz.
b. 6.35 mm thick

Appendix B
SI Conversion Factors

TABLE B-1 CONVERSION FACTORS AND SYMBOLS OF MAGNITUDE FOR SI UNITS

Conversion factors

To convert SI	to	multiply by
density, Mg/m^3	g/cm^3	1
	$lb/in.^3$	27.70
heat capacity, $J/kg \cdot K$	$cal/g \cdot °C$	2.39×10^{-4}
	$Btu/lb \cdot °F$	2.39×10^{-4}
permeability, darcy	m^2	9.869×10^{-13}
stress, Pa	kg/mm^2	1.02×10^{-7}
	$lb/in.^2$	1.45×10^{-4}
thermal conductivity, $W/m \cdot K$	$cal/cm \cdot s \cdot °C$	2.39×10^{-3}
	$Btu \cdot in./ft^2 \cdot h \cdot °F$	6.9396

Symbols of magnitude

Prefix	Symbol	Factor
giga	G	10^9
mega	M	10^6
kilo	K	10^3
milli	m	10^{-3}
micro	μ	10^{-6}

Other Noyes Publications

ULTRASTRUCTURE PROCESSING OF ADVANCED STRUCTURAL AND ELECTRONIC MATERIALS

Edited by

L.L. Hench
Department of Materials Science and Engineering
University of Florida

This book presents studies in the science of ultrastructure processing. As used here, ultrastructure processing means the manipulation and control of surfaces and interfaces to obtain new, high performance materials with predictable properties and environmental insensitivity.

Materials areas that may benefit from the use of ultrastructural processing include: particulate solids behavior, adhesion of fillers and reinforcers in composites, corrosion of glasses and glass-ceramics, fatigue of brittle materials, grain boundary attack of ceramics, effects of energetic particle beams, lifetime of non-oxide ceramics, electronic behavior of high band gap semiconductors, and multiphase electronic components.

Experimental approaches to ultrastructure processing described are:

(1) A systematic investigation of environment-surface interactions of silicate glasses and glass-ceramics, and non-oxide ceramics;

(2) the production of unique ceramics, glasses, glass-ceramics, and composites by the use of "transformation processing" (producing materials directly from chemical conversions rather than traditionally forming materials by compacting and densifying large particulates into objects); and

(3) "micromorphology processing"—production of submicron spherical powders, control of powder surface chemistry, and controlled assembly of particulates into ceramic or composite bodies.

A **condensed table of contents** is listed below.

1. INTRODUCTION—OVERVIEW

2. GLASS SURFACES

3. HYDROTHERMAL CORROSION OF LITHIA DISILICATE GLASS-CERAMICS

4. FRACTURE MECHANICS AND FAILURE PREDICTIONS FOR $Li_2O \cdot 2SiO_2$ GLASS AND GLASS-

CERAMICS

5. PREPARATION OF $xNa_2O\text{-}(1-x)\text{-}SiO_2$ GELS FOR THE GEL-GLASS PROCESS: I. ATMOSPHERIC EFFECT ON THE STRUCTURAL EVOLUTION OF THE GELS

6. PREPARATION OF $xNa_2O\text{-}(1-x)\text{-}SiO_2$ GELS FOR THE GEL-GLASS PROCESS: II. THE GEL-GLASS CONVERSION

7. PROTONIC CONDUCTON IN ALKALINE EARTH METAPHOSPHATE GLASSES CONTAINING WATER

8. PHOTO- AND THERMO-COLORING OF REDUCED PHOSPHATE GLASSES

9. DETERMINATION OF COMBINED WATER IN GLASSES

10. NOISE IN N-TYPE α-SILICON CARBIDE

11. LOW TEMPERATURE OXIDATION OF SiC

12. INTERGRANULAR SEGREGATION OF BORON IN SINTERED SILICON CARBIDE

13. MECHANISMS OF ELECTRON STIMULATED DESORPTION FROM SODA-SILICA GLASS SURFACES

14. SURFACE CHEMISTRY OF OXIDES IN WATER

15. SILICON NITRIDE AND SILICON CARBIDE FROM ORGANO-METALLIC AND VAPOR PRECURSORS

16. COMPATIBILITY OF A RANDOM COPOLYMER OF VARYING COMPOSITION WITH EACH HOMOPOLYMER

ISBN 0-8155-1004-7 (1984)

324 pages

Other Noyes Publications

FRACTURE IN CERAMIC MATERIALS
Toughening Mechanisms, Machining Damage, Shock

Edited by
A.G. Evans

Department of Materials Science and Mineral Engineering
University of California, Berkeley

This book presents recent studies on the mechanisms of fracture in ceramic materials—the effects of toughening, machining, and shock. Research on toughening mechanisms, machining and surface damage, thermal shock and general aspects of fracture in ceramic materials is described. Quantitative models of the various fracture processes have been developed. Special emphasis has been placed on the toughening that occurs in the presence of microcracks.

During the last decade, research on the fracture of monolithic single phase and multiphase ceramic polycrystals has attained a maturity which now permits many fracture phenomena to be quantitatively described. Specifically, the predominant fracture initiating flaws have been identified and the fundamental mechanics and statistics related to their fracture severity have been determined. In addition, the crack growth resistance exhibited by common ceramic microstructures can now be expressed in quantitative terms, through the development of micromechanics models of transformation toughening, microcrack toughening, and deflection toughening.

As a result, the next research frontier in the field of advanced monolithic ceramics undoubtedly resides in studies of the processing of optimum microstructures. Progress in this area is summarized in this book. The four main subject areas of the book are toughness/microstructure interactions, machining damage, thermal fracture and reliability, and impact damage.

A condensed table of contents listing **part and chapter titles** is given below.

I. TOUGHNESS/MICROSTRUCTURE INTERACTIONS

1. TOUGHENING MECHANISMS IN ZIRCONIA ALLOYS
2. THE MECHANICAL BEHAVIOR OF ALUMINA: A MODEL ANISOTROPIC BRITTLE SOLID
3. OBSERVATIONS OF INTERGRANULAR, CRACK DEFLECTION TOUGHENING MECHANISMS IN SILICON CARBIDE
4. ON THE CRACK GROWTH RESISTANCE OF MICROCRACKING BRITTLE MATERIALS
5. MICROSTRUCTURAL RESIDUAL STRESSES
6. SPONTANEOUS MICROFRACTURE IN MICROSTRUCTURAL RESIDUAL STRESSES
7. INDUCED MICROCRACKING: EFFECTS OF APPLIED STRESS

II. MACHINING DAMAGE

8. FAILURE FROM SURFACE FLAWS
9. SURFACE FLAWS IN GLASS
10. MECHANISMS OF FAILURE FROM SURFACE FLAWS IN MIXED MODE LOADING
11. GEOMETRICAL EFFECTS IN ELASTIC/PLASTIC INDENTATION
12. RESIDUAL STRESSES IN MACHINED CERAMIC SURFACES
13. FATIGUE STRENGTH OF GLASS: A CONTROLLED FLAW STUDY

III. THERMAL FRACTURE AND RELIABILITY

14. THE THERMAL FRACTURE OF ALUMINA
15. ASPECTS OF THE RELIABILITY OF CERAMICS FOR ENGINE APPLICATIONS

IV. IMPACT DAMAGE

16. LENGTH OF MAXIMAL IMPACT DAMAGE CRACKS AS A FUNCTION OF IMPACT VELOCITY

ISBN 0-8155-1005-5 (1984)

420 pages